T0306730

DoD Cyber Excepted Service Labor Market Analysis and Options for Use of Compensation Flexibilities

DAVID KNAPP, SINA BEAGHLEY, TROY D. SMITH,
MOLLY F. MCINTOSH, KAREN SCHWINDT, NORAH GRIFFIN,
DANIEL SCHWAM, HANNA HOOVER

Prepared for the Office of the Secretary of Defense
Approved for public release; distribution unlimited

RAND NATIONAL DEFENSE RESEARCH INSTITUTE

For more information on this publication, visit **www.rand.org/t/RRA730-1**.

About RAND

Research Integrity

Published by the RAND Corporation, Santa Monica, Calif.

Library of Congress Cataloging-in-Publication Data is available for this publication.
ISBN: 978-1-9774-0625-5

Limited Print and Electronic Distribution Rights

Preface

This report contains an analysis of the labor demand and supply of seven U.S. Department of Defense (DoD) cyber work roles that were collectively identified as high priority by the service components and the Office of the DoD Chief Information Officer (CIO). This analysis informs the use of compensation flexibilities granted to the DoD for its cyber workforce by the U.S. Congress and creates a potential approach for use of these compensation flexibilities. We provide a framework (based on economic theory) for adjusting pay, identify private-sector occupational counterparts for the seven work roles, discuss findings from DoD employment and compensation questionnaires completed by Cyber Excepted Service (CES) organizations, compare characteristics and life-cycle pay between DoD cyber civilians and their private-sector counterparts, and make recommendations aimed at supporting the DoD CIO in setting compensation policy.

Our analysis took place between April 2019 and July 2020 and required the collaboration and support from offices across the DoD. It is part of an ongoing study to evaluate compensation strategies for the CES. This report is written for a broad audience but is targeted at the Office of the DoD CIO and readers with a general background in the cyber field and personnel retention behavior of the federal civil service in the DoD. The research reported here was completed in November 2020 and underwent security review with the sponsor and the Defense Office of Prepublication and Security Review before public release.

This research was sponsored by the Deputy Chief Information Officer for Cybersecurity and conducted within the Forces and Resources Policy Center and the Cyber and Intelligence Policy Center of the RAND National Security Research Division (NSRD), which operates the National Defense Research Institute (NDRI), a federally funded research and development center sponsored by the Office of the Secretary of Defense, the Joint Staff, the Unified Combatant Commands, the Navy, the Marine Corps, the defense agencies, and the defense intelligence enterprise.

For more information on the RAND Forces and Resources Policy Center, see www.rand.org/nsrd/frp or contact the director (contact information is provided on the webpage). For more information on the RAND Cyber and Intelligence Policy Center, see www.rand.org/nsrd/intel or contact the director (contact information is provided on the webpage).

Contents

Figures

Tables

Summary

In 2016, Congress created the Cyber Excepted Service (CES) and granted the U.S. Department of Defense (DoD) flexibilities in setting compensation that were aimed at supporting the recruitment and retention of personnel critical to the DoD cyber warfare mission. The DoD Chief Information Officer (CIO), in conjunction with the Under Secretary of Defense for Personnel and Readiness (USD[P&R]), can set base compensation for the CES hire and "retain a quality workforce to execute the Defense Cyberspace mission" (Department of Defense Instruction [DoDI] 1400.25, Vol. 3006, 2017, p. 10). In this report, we analyze the labor demand and supply of seven DoD cyber work roles collectively identified as high priority by the service components and the Office of the DoD CIO. This report provides a cyber labor market analysis that is intended to inform the use of compensation flexibilities permitted under the CES for these work roles and provide a potential approach for the future use of compensation flexibilities.

The project team reviewed key components of compensation in the CES and identified that most compensation flexibilities outside of base compensation (e.g., cash incentives, alternative work schedules) are controlled by the employing federal agencies, rather than the CES, or are standard for federal employees (e.g., retirement benefits, group insurance). One compensation flexibility granted to the CES is the use of a targeted local market supplement (TLMS), which is an adjustment to permanent pay that can be narrowly defined to reflect a specific cyber work role, grade, and location. To justify a market-based permanent pay adjustment, there should be evidence that existing compensation is insufficient to attract and retain a required number of qualified employees. This can be identified by a persistent labor shortage, which can be demonstrated through high employee turnover or difficulty filling posted vacancies. Employers should first try temporary pay adjustments, such as cash recruiting and retention incentives, to solve the shortage. If the labor shortage persists, then permanent pay adjustments might be required; understanding existing pay discrepancies can help employers identify where large (i.e., structural) pay adjustments are needed. As the agents responsible for establishing TLMSs and other market-based compensation rate ranges, the DoD CIO and USD(P&R) need to document cyber work role shortages in CES-covered agencies. They also need to demonstrate that existing compensation flexibilities—specifically, recruiting, relocation, and retention incentives—are insufficient (DoDI 1400.25, Vol. 3006, 2017, pp. 9–10).

To identify cyber labor shortages and document the use of existing compensation flexibilities to address those shortages, the Office of the DoD CIO developed an employment and compensation questionnaire with input from the RAND project team and sent it to human resources offices in organizations that have converted or will convert to the CES in the near future. We reviewed responses provided by these organizations about employment, vacancies, turnover, and compensation.

The responses provided were generally incomplete and inconsistent. Collectively, the data provided by the services were insufficient to determine whether labor shortages exist in the CES workforce. However, specific cases were identified in which evidence suggests high turnover and substantial vacancies indicative of a labor shortage (e.g., Cyber Operators in U.S. Marine Corps Forces Cyberspace Command, Authorizing Official/Designating Representative in Navy's U.S. Fleet Cyber Command) and cases in which there were substantial differences in pay (e.g., cyber civilians in the Seattle local pay area). In these cases, the use of compensation flexibilities might be warranted. Although we considered only market-based reasons for permanent pay adjustments, other reasons might exist, including compensation equity or mission risk. We also reviewed civilian compensation data and found limited use of recruiting and retention incentives for workers in cyber work roles between 2010 and 2018. In contrast, performance awards were frequently used.

When an employer is experiencing a labor shortage that reflects pay insufficiency (as opposed to administrative difficulties in job posting and hiring), comparing this employer's workforce with workers at similar employers might reveal reasons for these difficulties and provide a reference point for adjusting compensation. We found that DoD civilian cyber workers are older and less likely to have a college degree; therefore, they potentially have more years of experience than private-sector cyber workers. DoD civilian cyber workers also are more likely to be U.S. citizens and to be veterans than are private-sector cyber workers. We find gender representation is similar across the two workforces. We also find little divergence in average weekly work hours.

When comparing compensation among DoD civilian and private-sector cyber workers, we primarily focus on the Washington, D.C., local pay area because it represents the greatest concentration of the DoD civilian cyber workforce (20 percent). We find some similarities and some differences in pay trajectories across the cyber work roles. In all cases, there is a DoD civilian pay premium at hiring. It is largest for Authorizing Official/Designating Representative—roughly $17,000—and smallest for Cyber Operator—roughly $6,000. However, that DoD civilian pay premium shrinks as years of potential experience increase. For all but one work role (Authorizing Official/Designating Representative), a private-sector pay premium emerges. The DoD civilian pay premium vanishes earliest for the Cyber Operator career (at ten years of potential experience). The private-sector pay premium that emerges midcareer for Cyber Defense Analyst, Security Control Assessor, Systems Security Analyst, and Cyber Defense Incident Responder is small (roughly $3,000) and remains small after 20 years of potential experience. In contrast, the private-sector pay premiums that emerge mid-career for Software Developer and Cyber Operator are large (roughly $12,000 and $20,000, respectively), and they persist through 30 years of potential experience.

The project team makes four recommendations. First, we recommend that the DoD continue to categorize cyber personnel by cyber work roles. This is necessary for facilitating analysis of the positions and determining whether current compensation is sufficient to fill them. Second, we recommend that the Office of the DoD CIO regularly collect data on DoD and private-sector employment and compensation of cyber positions through an annual CES employer survey. Regularly collecting this information will facilitate the DoD CIO in identifying persistent labor shortages and tracking use of recruiting and retention incentives. Third, we recommend the DoD CIO commit to a TLMS adjustment schedule over five years, based on verifiable hiring and retention benchmarks collected through administrative data, and communicate this to the organizations covered by this personnel system and their workforces. The intent of this approach would be to provide an incentive for these organizations to collect and track the information necessary for the Office of the DoD CIO to justify use of the compensation flexibilities it has been granted. Finally, we recommend that the Office of the DoD CIO consider structural pay adjustments using a TLMS for cyber work roles with existing labor shortages and major salary differences. In the Washington, D.C., local pay area, major salary differences were limited to Cyber Operators.

A limitation of our analysis is that we consider only market-based reasons for permanent compensation adjustments. It was outside the scope of our analysis to consider issues of compensation equity with other federal cyber employers (e.g., the Intelligence Community) or issues associated with mission risk (e.g., insufficient or unqualified personnel, combined with imminent need). These may provide separate, non–market-based rationales for permanent pay adjustments.

Acknowledgments

We would like to thank Patrick Johnson and Walter Spears from the Office of the U.S. Department of Defense Chief Information Officer for their insights, comments, and support during this portion of the study. At the RAND Corporation, we appreciate the depth of knowledge provided by Lily Ablon and the five cyber subject-matter experts who contributed to the project. We also appreciate support from Chris Broecker, Bobbie Sanders, Matt Isnor, and Janese Jackson from the Office of the U.S. Department of Defense Chief Information Officer. Our research benefited from the review of Kathleen Mullen, Ellen Tunstall, and Craig Bond.

Abbreviations

ACS	American Community Survey
AFCYBER	Air Force Cyber
ARCYBER	Army Cyber
BLS	U.S. Bureau of Labor Statistics
CES	Cyber Excepted Service
CIO	Chief Information Officer
CMF	Civilian Master File
CPF	Civilian Pay File
CPI-U	Consumer Price Index for All Urban Consumers
DMDC	Defense Manpower Data Center
DoD	U.S. Department of Defense
DoDI	Department of Defense Instruction
FLTCYBER	U.S. Navy's U.S. Fleet Cyber Command
FY	fiscal year
GAO	U.S. Government Accountability Office
GS	general schedule
KSAs	knowledge, skills, and abilities

KSATs	knowledge, skills, abilities, and tasks
JFHQ-DoDIN	Joint Force Headquarters—DoD Information Network
LPA	locality pay area
MARFORCYBER	U.S. Marine Corps Forces Cyberspace Command
NICE	National Initiative for Cybersecurity Education
NIST	National Institute of Standards and Technology
O*NET	Occupational Information Network
OES	Occupational Employment Statistics
OPM	U.S. Office of Personnel Management
PUMA	Public Use Microdata Area
PUMS	Public Use Microdata Sample
SME	subject-matter expert
SOC	Standard Occupational Classification
TLMS	targeted local market supplement
USCYBERCOM	U.S. Cyber Command
USD(P&R)	Under Secretary of Defense for Personnel and Readiness
YOPE	years of potential experience

CHAPTER ONE

Introduction

Congress created the Cyber Excepted Service (CES) in Section 1107 of the 2016 National Defense Authorization Act (Pub. L. 114-92, 2015). The authorities in this section were intended to attract and retain high-caliber personnel critical to the U.S. Department of Defense (DoD) cyber warfare mission. The separate excepted service was created because of an ongoing concern, both within Congress and among public policy organizations, that the U.S. government is not well positioned to compete in the market for cyber talent (Libicki, Senty, and Pollak, 2014; Partnership for Public Service and Booz Allen Hamilton, 2015; U.S. Government Accountability Office [GAO], 2011). Key challenges for the federal government to meet its cyber workforce needs include demand exceeding supply for cyber workers in the broader labor market, a skills gap in federal cyber positions, and agency workforce plans that do not address cyber workforce needs (Francis and Ginsberg, 2016).

In this report, we analyze the labor demand and supply of seven DoD cyber work roles that the service components and the Office of the DoD Chief Information Officer (CIO) collectively identified as high priority. A major motivation underlying the creation of the CES is that demand for cyber workers outstrips the supply. If true, then competition in the cyber labor market might make it difficult to fill and retain DoD cyber positions and jeopardize the DoD's ability to fulfill its cyber warfare mission. This analysis is intended to inform the use of compensation flexibilities permitted under the CES for these work roles; the approach can inform the use of compensation flexibilities for other cyber work roles.

One compensation flexibility granted to the CES is the use of a targeted local market supplement (TLMS), which is an adjustment to permanent pay that can be narrowly defined to reflect a specific cyber work role, grade, and location. To justify market-based permanent pay adjustment, there should be evidence that existing compensation is insufficient to attract and retain a required number of qualified employees. This can be identified by persistent labor shortage, which can be demonstrated through high employee turnover or difficulty in filling posted vacancies. Employers should first try temporary pay adjustments, such as cash recruiting and retention incentives, to close the shortage. If the labor shortage persists, then permanent pay adjustments might be required. Understanding existing pay discrepancies could help employers identify if and where large (i.e., structural) pay adjustments are required.

As the agents responsible for establishing TLMS and other market-based compensation rate ranges, the DoD CIO and Under Secretary of Defense for Personnel and Readiness (USD[P&R]) need to document cyber work role shortages in CES-covered agencies. Further, they need to demonstrate that existing compensation flexibilities—specifically, recruiting, relocation, and retention incentives—are insufficient (Department of Defense Instruction [DoDI] 1400.25, Vol. 3006, 2017, pp. 9–10).

A limitation of our analysis is that we only consider market-based reasons for permanent compensation adjustments. It is outside the scope of our analysis to consider issues of compensation equity with other federal cyber employers (e.g., the Intelligence Community) or issues associated with mission risk (e.g., insufficient or unqualified personnel, combined with imminent need). These might provide separate, non–market-based rationales for permanent pay adjustments.

In this chapter, we briefly review key aspects of CES administration and the compensation flexibilities granted to it by Congress. We introduce a framework developed by the National Institute of Standards and Technology (NIST) for classifying cyber work roles and discuss its implications for making labor market comparisons. In addititon, we provide a framework, based on economic theory, for adjusting wages; this framework reflects challenges distinct to the DoD and the implications of setting compensation policy. In Chapter Two, we review different forms of compensation and benefits available to federal and private-sector workers, highlighting key differences in pay and benefits that might influence recruiting and retention outcomes. In Chapter Three, we discuss existing measurements of demand for cyber work roles among CES organizations. In Chapter Four, we discuss how we identify private-sector occupational counterparts for the seven DoD cyber work roles we analyze. In Chapter Five, we report on differences in characteristics and compensation of DoD civilian and private-sector cyber workers (i.e., labor supply). In Chapter Six, we summarize our findings, discuss the implications of these findings for setting compensation policy, and provide recommendations for how to approach initially setting and subsequently updating CES compensation.

CES Administration and Flexibilities

The CES is administered by the DoD CIO, in conjunction with the Office of the USD(P&R). DoD employees in "CES positions must perform, manage, supervise, or support functions necessary to execute the responsibilities of the United States Cyber Command" (DoDI 1400.25, Vol. 3001, 2017). Although the DoD CIO administers CES compensation policy, it does not direct human resources actions or budgets of DoD organizations with CES employees, including the use of recruiting, retention, and relocation incentives.

The law creating the CES provides for hiring and compensation flexibilities. The CES is a part of the excepted service, which is one of the three federal government hiring authorities: competitive service, excepted service, and senior executive service. The majority of positions in the DoD cyber workforce prior to the CES were competitive service positions on the general schedule (GS). In the competitive service, applicants undergo a competitive examining process that might include a written test, an evaluation of the applicant's experience and education, and an evaluation of other attributes deemed necessary to perform the position (U.S. Office of Personnel Management [OPM], undated).[1] Veterans' preference is applied as part of this process.

The CES, as an excepted service, does not apply OPM's competitive examining process but applies its own employment policy (DoDI 1400.25, Vol. 3005, 2017). CES employment policy is intended to adhere to merit-based principles. Recruitment is not limited by requirements for public notification or vacancy notices and is intended to be fully flexible (i.e., no requirements that certain positions be opened or closed to particular candidate groups, such as internal candidates). In evaluating position qualifications, the CES can apply its own qualification standards. Qualification standards for internal hires do not impose time-in-grade requirements, which are otherwise common in civil service positions.

The CES also applies its own compensation policy (DoDI 1400.25, Vol. 3006, 2017). The DoD CIO, in conjunction with the USD(P&R), sets CES compensation policy. As of July 2020, CES compensation mirrors the GS pay system. Positions in the CES are on the General Government pay schedule; current policy is for this schedule to automatically reflect changes to the GS base pay schedule. A difference is that the CES General Government schedule offers two pay steps in addition to the standard ten pay steps per pay grade. CES compensation includes local market supplements, which apply uniformly to all CES employees in a pay area and mirror GS locality rates. An additional flexibility available to the CES is the TLMS, which replaces the local market supplement and can be applied in response to labor market conditions that are not fully addressed by the CES pay grade and local market supplements. Base compensation, including locality pay, is limited to Level IV of the Executive Schedule (which was $170,800 in 2020). Base compensation does not include incentives and allowances. Aggregate annual compensation cannot exceed Level I of the Executive Schedule (which was $219,200 in 2020).

---

1 OPM's qualification standards typically set education and experience requirements for broad groups of jobs (e.g., professional and scientific positions) and supplementary requirements for specific occupations (e.g., GS-1550, computer scientist: at least 15 of the 30 semester hours must have included any combination of statistics and mathematics that included differential and integral calculus).

CES compensation policy also allows for the creation of "new CES base compensation rate ranges applicable to specific DoD Components, locations, occupational groups, and specialties" (DoDI 1400.25, Vol. 3006, 2017, p. 10). Changing base compensation is only permitted when the market situation for the specific workforce group "is such that separate policy considerations are deemed necessary to maintain the integrity of the CES compensation framework" (DoDI 1400.25, Vol. 3006, 2017, p. 10). Justification requires

> detailed analysis of recruiting or retention issues regarding the targeted occupational or specialty groups. It also requires supporting evidence that other actions within the existing CES policy framework, including recruitment, relocation, and retention incentives, are insufficient to ensure successful maintenance of the required workforce. (DoDI 1400.25, Vol. 3006, 2017, p. 10)

Moreover, the establishment of these rate ranges will be supported by "ongoing review of the effectiveness of the new base compensation structure and trigger indicators for phasing the new structure into the core CES framework as conditions warrant" (DoDI 1400.25, Vol. 3006, 2017, p. 10).

Categorizing Cyber Work Roles

In the late 2000s and early 2010s, government leaders, including the Federal Chief Information Officers Council, recognized that the federal cybersecurity workforce had not been defined or assessed. This spurred the creation of the National Initiative for Cybersecurity Education (NICE) workforce framework. The NICE framework, which is led by the NIST, provides consistent classification for cybersecurity work in the public, private, and academic sectors, establishing a set of work roles and required knowledge, skills, abilities, and tasks (KSATs; Newhouse et al., 2017) for cybersecurity and related work, irrespective of where or for whom the work is being performed. The first version of the framework was released in September 2011, and subsequent versions adopted input from the Department of Homeland Security and the Office of the Secretary of Defense.

The DoD adopted the NICE framework for classifying CES positions (DoDI 1400.25, Vol. 3006, 2017, p. 4). The NICE classification of positions exists independently of and concurrently with OPM's occupational series. The DoD CIO has classified all filled DoD cyber positions with NICE work roles.[2] Using this classification, we find that most cyber work roles align with certain occupations in OPM's occupation series, including information technology management (2210), computer science (1550), and miscellaneous administration and program (0301).

Classifying positions within a framework, such as the NICE framework, assists in the ability to identify, recruit, and develop cybersecurity workers (Newhouse et al., 2017). However, this classification has not been adopted by major public labor market surveys, such as the Current Population Survey or the American Community Survey (ACS). Instead, these surveys use the Department of Labor's Standard Occupational Classification (SOC) system. There is no standard crosswalk between NICE work roles and SOC occupations. However, to compare the DoD civilian and private-sector cyber workforces, we need a crosswalk. In Chapter Four, we use NICE KSATs to identify the best possible matches, and we use these matches when comparing DoD civilian and private-sector cyber worker characteristics and earnings in Chapter Five.

Theoretical Framework for Potential DoD CES Compensation Changes

The DoD's ability to efficiently maintain its required cyber workforce depends on characteristics of the cyber worker labor market. A

---

[2] A GAO report (2019) found work roles were classified incorrectly and inconsistently. We used position classifications that were completed by the the Office of the DOD CIO in early 2019, after the GAO report. As part of the more recent classification, the Office of the DoD CIO made efforts to improve the classification process, but we did not independently confirm classifications.

labor market comprises employers that demand labor and workers who supply labor. A labor market is in *equilibrium* when workers' compensation is sufficient, meaning that there is not excess demand for labor by employers (i.e., there is no labor shortage) and there is not an excess supply of labor (i.e., there is no labor surplus). A key motivation for an employer, such as the DoD, is to offer the lowest wage to keep its positions filled with qualified workers, relying on existing compensation, promotion pathways, and incentives.[3]

Employment dynamics are important when an employer is deciding whether to adjust wages. If a position is critical to the employer's output, as cyber talent is to the U.S. Cyber Command (USCYBERCOM), then position vacancies must be posted and filled in a timely manner. Persistent vacancies and high turnover represent the strongest case for adjusting pay. Benchmarks for filling vacancies and limiting turnover should be established to determine when compensation adjustments are needed to address or prevent labor shortages.

How much to adjust pay and how quickly depends on the necessity of the position to the employer's output (e.g., for the DoD CES, a cyber civilian position's contribution to the U.S. cyber warfare mission) and alternatives to filling that position. For example, a cyber civilian's position that is difficult to fill could instead be filled by a cyber service member or a contractor with similar knowledge, skills, and abilities (KSAs). It could also be filled by a cyber civilian with similar KSAs but in a different location. A position also might be replaced or modified with the introduction of new technology. Finally, the tasks associated with a position might be reallocated among existing employees.

Each of these choices has an opportunity cost, some of which are easily measured (e.g., a contractor's salary, the cost of new technology) and some of which are difficult to measure (e.g., the inability to use a service member in other capacities, work going undone). The labor

3 We consider the DoD as a cyber employer, but DoD components could be thought of as multiple employers competing in the labor market for cyber talent. This can introduce additional complexities that we do not explore here. The philosophy of the CES is to coordinate across the DoD components to set a compensation policy that "supports their individual and collective organizational mission, goals, and objectives" (DoDI 1400.25, Vol. 3006, 2017, p. 22).

market is a spot market, meaning that positions in need of labor will be filled by workers in need of employment at that point in time. If a position is an immediate need for an employer, then the employer would want to increase compensation to entice potential employees. Higher compensation should be offered for more-critical positions and for positions that have few feasible alternatives. If a position is not immediately needed, an employer could continue searching for a qualified applicant who is willing to accept the position at the offered wage.

Pay also depends on which sector of the economy an employer is in and whether the employer is a government entity, a private for-profit firm, or a nonprofit organization. Traditionally, base salaries and other monetary incentives have been lower in the federal government and nonprofits than in private-sector jobs. Falk (2017, p. 13) found that the highest salaries available in the federal government were "substantially lower than the average salaries for most executive positions in the private sector." However, a Congressional Budget Office study noted that, for some education levels, federal employees earn higher wages than private-sector employees (Falk, 2017). For employees with a bachelor's degree or less, a federal employee could make an equivalent or even larger wage per hour than in the private sector. Conversely, employees with advanced professional degrees often receive higher wages in the private sector than in the federal government (Falk, 2017). Other research has revealed that, although the average cybersecurity professional's salary falls within the range of the federal government's pay schedules ($80,000 to $100,000), upper-tier cyber professionals can sometimes make more than $250,000 per year, which the government might have difficulty providing (Libicki, Senty, and Pollak, 2014).

A common misconception is that pay differences with outside employment options are sufficient to merit a pay adjustment. A new worker's decision to join a firm and a current worker's decision to stay at a firm depend not just on pay, but on other forms of measurable compensation (such as those discussed in Chapter Two) and harder-to-measure compensating differentials. Examples of compensating differentials include a worker's preferences for a certain type of employment (e.g., performing public service) or job security. A position that offers other forms of compensation or positive compensating differentials could attract a worker at comparatively lower pay. Therefore, an efficient compensation system is based on market signals, not on pay differences alone.

DoD wages depend primarily on a position's pay schedule and grade, but the organization has some latitude in setting pay, as discussed in the next chapter. The DoD CIO is able to make a business case demonstrating the need for higher base pay for a position in a CES-covered organization that has persistent difficulty in filling a necessary CES position. This higher pay could come in the form of a TLMS or an alternative pay schedule. Such changes should be regarded as persistent policy changes because base wage changes are rarely reversed (Fallick, Villar, and Wascher, 2020). Either option should be based on (1) an analysis of recruiting or retention issues for that position and (2) supporting evidence that other actions within the existing CES policy framework, including recruitment, relocation, and retention incentives, are insufficient. Such a business case should consider the employer's need, including necessity and time horizon, for that position. It should also consider the effect of the additional cost of paying more to employees who do not require the higher pay to stay.

CHAPTER TWO

Comparison of Federal Civilian and Private-Sector Compensation

Compensation plays an important role in recruiting and retaining employees; it is the "cornerstone for recruiting, retaining, and motivating the type of employees" needed for DoD cyber components (DoDI 1400.25, Vol. 3006, 2017, p. 22). Compensation packages vary across occupations and employers, but a standard set of components is found in most compensation packages. This includes monetary compensation (e.g., base pay, allowances, incentives, loan repayment, retirement savings) and nonmonetary compensation (e.g., access to group insurance, work flexibilities, paid leave, training opportunities).

In this chapter, we review standard compensation components that are available to the federal civilian and private-sector workforces and highlight types of compensation that might best influence recruiting and retention outcomes.[1] We pay particular attention to types of federal compensation that could be leveraged to attract and retain workers in the CES, as well as what alternative types of compensation DoD might require to remain competitive. To identify the types of compensation available in the federal government and private sector, we reviewed academic literature on compensation packages, employee and employer surveys on compensation, blogs on recent compensation trends, OPM handbooks on compensation policies, technology and cyber workforce compensation surveys, and publicly available infor-

---

[1] We use *types of compensation* to reflect specific compensation options, typically within the context of a compensation component. For example, many jobs will have retirement benefits, a component of compensation. A type of retirement benefit would be a defined benefit pension plan (uncommon in the private sector, but available for federal workers).

mation on compensation packages offered by major tech companies (specifically, Amazon, Apple, Facebook, Netflix, and Google) and by smaller cybersecurity companies located in Washington, D.C.[2]

Although there is considerable overlap in the types of compensation offered by the federal government and the private sector, there are some differences. These similarities and differences may guide the DoD CIO in identifying which types of compensation they can leverage to recruit and retain the CES workforce.

Components of Compensation

We identified the following eight standard components of compensation: (1) base compensation, (2) incentives, (3) allowances, (4) group insurance, (5) retirement benefits, (6) leave opportunities, (7) work flexibilities and perks, and (8) training and development. Each component consists of multiple types of compensation. Not all types of compensation are exclusive to a single component. For example, paid time off is a part of most compensation packages. In the federal government, it sometimes also is a performance incentive.

Table 2.1 shows the eight components of compensation available in the federal government and the private sector and provides examples of associated types of compensation. For each of the eight components, we briefly describe the characteristics of that compensation component and highlight how types of compensation differ across employers (e.g., for group insurance, we describe differences in copays and contributions). As most types of compensation are available for federal and private-sector workers, we emphasize when a type of compensation is not available or uncommon in one of these sectors or if there are differences by sector use.

---

[2] Although compensation surveys, which we will refer to as *trade surveys*, provide insights into recent trends in compensation, there is limited information on the methodologies used to conduct these surveys, and most appear to reflect convenience samples. For this reason, we do not report statistics from these surveys, but we do use them to understand the types of compensation offered.

Table 2.1
Components and Types of Compensation

Components of Compensation	Types of Compensation
Base compensation	• Salary or hourly wage • Commissions • Tips • Stock options • Minimum bonus or profit-sharing
Incentives	• Signing bonus/recruitment incentive • Retention incentive • Relocation incentive • Bonus/profit-sharing/other performance incentive • Awards and recognition • Referral bonus • Premium pay (e.g., overtime, hazard) • Paid time off
Allowances	• Transportation assistance • Cost-of-living allowance • Travel allowance • Meal allowance • Housing allowance
Group insurance	• Health, dental, and vision insurance • Life insurance and accidental death and dismemberment insurance • Short- and long-term disability insurance • Long-term care insurance • Professional liability insurance • Travel accident insurance
Retirement benefits	• Defined benefit pension account • Defined contribution pension account • Retiree health, dental, and vision insurance • Retiree life and accidental death and dismemberment insurance
Paid Leave	• Vacation • Sick leave • Paid or personal time off (combines vacation, sick, etc.) • Holidays • Family leave • Miscellaneous paid leave: bereavement, jury duty, military service, personal days

Table 2.1—Continued

Components of Compensation	Types of Compensation
Work flexibilities and perks	• Teleworking/remote work • Alternative work schedules • Subsidized child care • Flexible spending accounts • Health and wellness program • Employee assistance program • Personal use of firm resources • Food and drinks at workplace • Discounts on products and services
Training and development	• Training opportunities • Developmental opportunities • Tuition assistance or reimbursement • Loan repayment

NOTES: The classification of compensation into components and types of compensation was compiled by the authors based on a review of academic, government, and trade literature in the human resources field, including Dice, 2019; Harvey, 2018; OPM, 2013; Payscale, 2019; and Taras, 2012.

Base Compensation

Base compensation typically comes in the form of a salary or an hourly wage. Federal employees in the CES are salaried employees. There are different ways to structure base salaries, including pay range and grade-step structures. The "2018 Salary and Structure Policies and Practices Survey," a trade survey based on a convenience sample, indicates *pay ranges* are the most popular pay structure. Pay ranges provide a range or band of pay for a particular position or a common group of jobs and impose limited structure on where salaries are set within the range or band (Cybulski, Sever, and Stoskopf, 2019). Examples of pay band systems within the federal government include the Acquisition and Laboratory Demonstration projects and the Senior Executive Service. However, the federal government primarily uses a *grade-step structure.* In a grade-step structure, a position is assigned a grade; there are steps within a grade, with higher steps earning higher pay. An employee can progress to higher steps with additional tenure or education.

In addition to salary, base compensation may include commissions and tips, which we believe are uncommon in either the federal or private-sector cyber workforce. Stock options and guaranteed minimum levels of profit-sharing or bonuses, which exist in the private sector but not in the federal government, provide a means of encouraging sustained high performance among employees because their future compensation, through firm stock or profit, depends on their current effort.

Incentives

Monetary incentives include recruiting, relocation, and retention incentives; referral bonuses; and performance-based bonuses. These incentives can be powerful tools for recruiting and retaining employees when employers face restrictions on base salaries. An important advantage of incentives is that they generally are not permanent. The federal government and the private sector use incentives, although the federal government has limitations on the size of these incentives (OPM, 2013). For instance, recruitment incentives "may not exceed 25 percent of the employee's annual rate of basic pay in effect at the beginning of the service period multiplied by the number of years in the service period" (OPM, 2013, pp. 41–42). Relocation and retention incentives have similar restrictions. In certain situations, OPM may approve recruitment and relocation incentives above this limitation.

Nonmonetary incentives include additional paid time off and awards and recognition. We found that additional paid time off is an incentive used in the federal government, but it was not referenced in any of the nongovernment material that we reviewed. Profit-sharing is not available in the federal government, but most of the other types of incentives listed in Table 2.1 exist in some form for federal workers.

Allowances

Allowances are maximum permissible amounts that an employee may use for a work-related requirement that would otherwise financially burden the employee. A typical allowance in urban areas is transportation assistance in the form of parking or money toward public transport fares.

We found that allowances as part of a compensation package (as opposed to allowances while on business travel) are not common. The exceptions are transportation assistance and cost-of-living allowances for overseas assignments, which exist for federal workers. Meal allowances, which are distinct from perks (such as free food and drinks at work sites), are typically associated with private-sector jobs that require on-site work beyond normal business hours.

Group Insurance

Group insurance policies are common employer benefits and are available to federal employees. A group insurance policy can be less expensive than buying insurance as an individual because the group often represents a less costly risk pool. Employers may pay a portion of group insurance premiums. For example, medical insurance is more likely to be partially paid by the employer than other forms of insurance, such as life, accidental death and dismemberment, or long-term care (U.S. Bureau of Labor Statistics [BLS], 2019). For employers with defined benefit pension plans, disability insurance is usually part of the pension plan. All federal workers have a defined benefit pension plan (see our later section on retirement benefits for more detail), but this type of plan is not used by most private-sector employers. Consequently, private employers typically offer their employees some form of short- and long-term disability insurance.

Insurance policies vary greatly, and (according to some research) health insurance policies at smaller private-sector organizations are often less generous than those of larger companies or the federal government (Falk, 2017). Generosity of insurance plans is generally determined by the size of the group insured, the amount an organization contributes toward the cost of insurance, the insurance plan's copay (i.e., payments when the insurance is used), maximum benefit amounts, and coverage.

Retirement Benefits

Retirement pension plans typically fall into two main categories: *defined benefit* and *defined contribution* plans. A defined benefit plan provides an annuity based on an employee's salary history and their years of service. It is guaranteed by the employer and generally includes provisions for situations in which the worker becomes disabled and for the worker's survivors should the worker die (BLS, 2020a). These benefit provisions typically extend into retirement. The federal government has two defined benefit pension plans: the Civil Service Retirement System, which includes people hired in and before 1986, and the Federal Employee Retirement System, which includes those hired after 1986.

In contrast, the value of a defined contribution plan is based on annual contributions and investment earnings. The investment earnings are managed by the employee, and the employee is responsible for the investment risk. Defined contribution plans differ by how much the employer contributes (Dulebohn et al., 2009; Falk, 2017). Federal employees covered under the Federal Employee Retirement System are eligible for an employer-matched defined contribution plan as well, called the Thrift Saving Plan, in which the federal government matches employee contributions up to 5 percent of the employee's salary.[3]

Defined benefit plans are less common in the private sector: Only 8 percent of private industry establishments have a defined benefit pension, compared with 49 percent that offer a defined contribution plan. In the professional and technical services, few firms offer a defined benefit plan, and 61 percent offer a defined contribution plan (BLS, 2019).

Beyond retirement pension plans, the federal government provides the option to extend some benefits into retirement, including most group insurances. This is rarely offered in the private sector: About 12 percent of private industry establishments offered retiree health insurance in 2019 (BLS, 2019).

Paid Leave

Paid leave policies include vacation or paid time off, sick leave, holidays, bereavement leave, military leave, jury duty leave, and family and medical leave (including maternity and paternity leave). Vacation or paid time off can take the form of accrued annual leave (i.e., leave accrued based on weeks worked), annual leave given as a lump sum at some point in the year, and unlimited time off (Payscale, 2019). The federal government uses an accrual policy that increments by years of service (OPM, 2013). The federal government also offers paid sick

---

3 The Thrift Savings Plan is also available to employees in the Civil Service Retirement System, but their contributions are not eligible for an employer match.

leave and holidays and, for births after October 1, 2020, 12 weeks of paid parental leave. (Public Law 116-92, 2019).

Paid leave is common in the private sector, but firms vary in the benefits offered. The 2019 National Compensation Survey indicates that more than 90 percent of establishments within the professional and technical services industry offered paid holidays, vacation, and sick leave.

Significant variation exists in paid family leave: Only 34 percent of workers in establishments within the professional and technical services industry offered paid family leave (BLS, 2019), although major tech companies (such as Facebook, Amazon, Apple, Netflix, and Google) offer more than 12 weeks of paid leave for primary caregivers; many also offer extended leave for secondary caregivers (Amazon, undated; Facebook, undated; Google, undated; Molla, 2018; Netflix, undated). Netflix offers its employees as much as a year of paid time off. Amazon has a "ramp-back" policy which allows parents to gradually return to work following the birth of a child. Amazon also has a unique leave-share program through which it will pay for up to six weeks of an employee's partner's salary if the partner does not have paid parental leave (Molla, 2018). While not all employers in the private sector offer such generous plans, some trade surveys suggest that paid family leave is becoming more common (Payscale, 2019).

Work Flexibilities and Perks

Compensation plans often include flexibilities in how and when an employee works and what perks are available to facilitate the ability to work and enjoyment in working. Key examples include teleworking and remote work, alternative work schedules, subsidized child care, flexible spending accounts, health and wellness programs, employee assistance programs, personal use of firm resources, food and drinks at workplace, and discounts on goods and services.

Work schedule and location flexibilities are becoming increasingly common throughout organizations. In 2020, many employers began offering more-flexible work schedules and remote-work policies following requirements to socially distance because of the coronavirus disease 2019 (COVID-19) pandemic (Aragon, 2020). For federal employees, these policies are at the discretion of the employing agency

(OPM, 2020). When flexible work schedules are permitted, they typically involve variable start and end times outside of the agency's regular work hours and shorter workweeks with longer days (OPM, 2013). Although federal agencies are technically able to offer flexible work hours and alternative work schedules, these still follow relatively strict rules. The potential sensitivity of federal work and geographic constraints reduce the federal government's ability to offer the type of flexible work schedules and remote-work opportunities that can be found in the private sector (Bedding and de Jongh, 2017). Some recent trade surveys show that more private-sector employers offer these types of options (Dice, 2019; Harvey, 2018; Payscale, 2019).

Child care assistance ranges from subsidized child care to on-site child care facilities. Many, but not all, federal agencies have access to on-site child care facilities (OPM, 2013). Health and wellness policies can include reimbursements for gym memberships, on-site fitness facilities, and other policies that encourage healthy living. Many employers, including federal agencies, also offer employee assistance programs, which provide counseling to employees on several issues, including substance abuse, family issues, and other challenges employees may be dealing with (OPM, 2013).

Other perks, including personal use of firm resources (e.g., cars) and food and drinks at the workplace, are more limited for federal employees. Some employers in the private sector offer on-site amenities. Several major tech companies offer on-site fitness centers, classes, cafes, laundry, and other services (Amazon, undated; Facebook, undated; Google, undated; Netflix, undated). No data were available on the prevalence of these types of perks among private-sector employers more broadly.

Training and Development

Continuous learning is important for a skilled cyber workforce to maintain professional certifications. Employers might provide training opportunities, reimburse employees for training and additional education, or reimburse student loans. OPM encourages federal agencies to create training programs, which can include tuition reimbursement, targeted training, and professional development opportunities (OPM, 2013). Agencies can also offer individual learning accounts that employees can use to pursue further development opportunities (OPM, 2013). In the federal government, student loan repayment can also be included in offer packages or as part of a retention incentive.

There is limited data on the inclusion of training and development opportunities as part of private-sector compensation packages. In one tech-focused worker trade survey, 71 percent of respondents stated that training and education opportunities were important to them (Dice, 2019). Some specific examples in major tech companies include Apple offering tuition reimbursement and Amazon's "Career Choice Program" that pre-pays the costs of educational opportunities for some employees (Amazon, undated; Apple, undated). Some employers have developed programs aimed at training new hires in the skills that the company requires as an alternative to hiring people who already have these skills. These bootcamp-style training courses teach skills, from coding to communication, that will be necessary for the new hires when they start their full-time positions as software developers or project managers (Gee, 2017). Some tech firms, including Google, IBM, Amazon, Salesforce, and Facebook, are partnering with universities and community colleges to offer credit-bearing certificate programs in information technology fields (Fain, 2019). Given that certifications are important qualification for cyber workers, we expect that this training and development support likely is salient to cyber workers; however, the frequency with which they are included in private-sector compensation packages is unknown.

Key Differences Between CES and Non-CES Federal Employee Compensation Policy

Federal compensation policy grants flexibilities to agencies to set compensation packages to respond to unique recruiting and retention needs while providing oversight to ensure these packages follow applicable policy and laws. As discussed in Chapter One, the CES offers some new compensation flexibilities, specifically (1) the option to add two steps onto a pay grade and (2) the ability of the DoD CIO, in conjunction with USD(P&R), to establish TLMSs for specific subpopulations.

The main innovation of the CES is that the responsibility for approving these flexibilities resides with the DoD CIO and USD(P&R) rather than with OPM. The intent of this shift is to facilitate greater responsiveness to DoD's cyber workforce recruiting and retention needs.

Otherwise, most of the compensation flexibilities afforded CES mirror those for the broader federal government. Most of these flexibilities are managed by the employing agency, including incentives, allowances, paid leave, work flexibilities, and training and development. Retirement benefits and group insurance are common across all agencies and are common between CES and non-CES federal workers.

Implications of Available Compensation Flexibilities for the CES

Many cyber employers compete with the CES for talent, and they all may provide different compensation packages. Consequently, the CES cannot establish one compensation package and expect that package to work in every case.

Most of the types of compensation discussed in this chapter are available in both the CES and private-sector cyber workforces. Key types of compensation that are available in the private sector but not to CES employees include stock options, profit-sharing, and such perks as personal use of firm resources (e.g., cars) or free food and drink at the workplace. Some common flexibilities are often unavailable for federal workers, such as work schedules and locations. Key types of compensation that are available to CES employees but are uncommon for private-sector employees include defined benefit pensions, retiree group insurances, access to long-term care insurance, additional paid time off as a performance award, and subsidized child care.

The federal government should highlight the compensation options available only in the federal government as part of the hiring process, perhaps to compensate for lower base compensation. However, it might be difficult to communicate the value of a form of compensation that is not common in the private sector. This is particularly true for deferred compensation, such as the defined benefit pension and access to retiree group insurance. Research suggests that a dollar of deferred compensation (e.g., future pension payments, contributions to retirement accounts) is valued less than a dollar of current compensation (Fitzpatrick, 2015; Goldhaber and Holden, 2018). However, deferred compensation is still valued, and federal employers could benefit from helping potential hires understand the value of these and other differentiating benefits.

If the DoD is having problems recruiting and retaining cyber workers, then the CES and affiliated agencies should take full advantage of the compensation flexibilities that are available to it, including pay-setting flexibilities and recruiting and retention incentives. Implementing more-flexible schedules, teleworking policies, and training and development opportunities could also attract potential employers. Although the federal government cannot offer the high salaries or perks seen in the private sector, there are other compensation options it can leverage.

In our review, we identified that most compensation flexibilities outside of base compensation are controlled by the employing federal agencies rather than the CES. Therefore, the DoD CIO, as administrator of the CES, can promote the use of compensation flexibilities, but it cannot direct their use. The DoD CIO, in conjunction with USD(P&R), can control base compensation for the CES and create policy around setting base compensation that establishes benchmarks for adjusting compensation that encourages agencies to efficiently use their compensation flexibilities (e.g., recruiting/retention incentives). This is a key point that we will return to in our recommendations in Chapter Six.

CHAPTER THREE

Cyber Work Role Shortages in CES-Covered Agencies

As discussed in Chapter One, a key motivation for a firm adjusting its wages is if its demand for labor is not met by the available supply of workers in the broader labor market. Documentation of persistent vacancies and high turnover of a position could demonstrate a shortage of labor and represent the strongest case for adjusting pay for that position.

Evidence suggests that the U.S. cyber labor market, including government and private-sector employers, is experiencing a shortage of cyber labor. Cyberseek.org publishes information that approximates the demand and supply for cyber workers.[1] Demand is measured by current cyber job postings, and supply is measured by employment in related occupations as captured in labor force surveys.[2] As of July 16, 2020, Cyberseek.org reported 504,316 U.S. job openings relative to 997,058 workers employed in cyber-related occupations: a ratio of two people employed for every one opening. A lower ratio is associated with a greater shortage of labor. The public sector (including federal, state, and local government) has a ratio of 1.7 people employed for every

1 Cyberseek.org is a partnership between NICE, the Computing Technology Industry Association, and Burning Glass Technologies. It is supported by NICE, a program of NIST in the U.S. Department of Commerce, under Grant #60NANB19D124. Burning Glass Technologies uses its own link between NICE work roles and SOC employment codes in providing the analysis available on Cyberseek.org.

2 Ideally, demand would measure position openings as well as existing positions. Supply would ideally reflect all workers qualified for and willing to work in cyber occupations, regardless of whether they were currently employed in a cyber position.

opening, suggesting that the public sector is experiencing a greater shortage of cyber labor.

To justify the use of its new compensation flexibilities, notably the TLMS, the DoD CIO needs to document cyber work role shortages in CES-covered agencies. CES compensation policy states that justification for use of these broad pay flexibilities should include

> detailed analysis of recruiting or retention issues regarding the targeted occupational or specialty groups, and supporting evidence that other actions within the existing CES policy framework, including recruitment, relocation, and retention incentives are insufficient to ensure successful maintenance of the required workforce (DoDI 1400.25, Vol. 3006, 2017, p. 10).

In addition, to understand demand for cyber positions, we need measures of labor supply and demand in CES-covered agencies. We worked with the Office of the DoD CIO to develop an employment and compensation questionnaire that the Office of the DoD CIO fielded to agencies either currently covered by the CES or that will be covered in the next few years. The intent of this questionnaire was to provide data to document the existence of recruiting or retention issues in these organizations and collect information on their use of existing compensation flexibilities.

In this chapter, we review the elements of that questionnaire and its key findings. We also present our analysis of independently collected DoD civilian pay data on the use of recruiting/relocation, retention, and performance incentives by these organizations. Finally, we discuss the implications of these findings for using CES compensation flexibilities.

Cyber Employment and Compensation Questionnaire

The CES is composed of cyber workers in several DoD agencies. Transition of civil service employees into the CES is being done by agency and includes all employees of those organizations, regardless of whether they are cyber workers. As of July 2020, the organizations that have converted include USCYBERCOM, Joint Force Headquarters—DoD Information Network (JFHQ-DoDIN), U.S. Marine Forces Cyberspace Command (MARFORCYBER), U.S. Navy's U.S. Fleet Cyber Command (FLTCYBER), and the Defense Information Security Agency. Air Force Cyber (AFCYBER) and Army Cyber (ARCYBER) will also convert in the coming years.

Human resources managers in these organizations were asked to complete the employment and compensation questionnaire. They were informed that the questionnaire responses would be used by the Office of the DoD CIO to better understand the characteristics of the CES workforce, identify ways to improve the compensation system to increase management flexibility, and develop an approach to support analysis of potential compensation policies aimed at retaining a high-quality cyber workforce. The questionnaire was fielded between May and October 2019.

The questionnaire included questions about employment, hiring, and turnover for the overall workforce, for cyber workers, and for specific cyber work roles. The following questions provide a sample of questions asked:

1. How many civilian employees are there in your organization?
2. How many vacant positions are there right now?
3. How many of these vacant positions will be filled over the next year?
4. How many employees have been separated in the last year?
5. How many job postings has your organization made this past year?
6. On average, how long does it take to receive approval for a new position posting?
7. How many offers have been made in the last year?
8. How many offers have been accepted?
9. Of the accepted offers, how many actually have been hired (tentative, firm, final, etc.)?
10. What is the average length of time between making an offer and start date?

The questionnaire also included questions about the agency's use of specific compensation flexibilities for the recruitment and retention of cyber workers. The following questions provide a sample of questions asked:

1. For each recruitment compensation flexibility listed below, please indicate whether your organization uses the flexibility and, if so, the percentage of your organization's new cyber job offers in the last year that include this benefit, as well as the average value of the benefit.
   a. recruitment incentive
   b. relocation incentive
   c. student loan repayment.

2. For each compensation flexibility for current cyber employees listed below, please indicate whether your organization uses the flexibility and, if so, the percentage of your organization's cyber employees participating or receiving this benefit in the last year and the average value of the benefit.
   a. retention incentive (likely to leave the federal service)
   b. retention incentive (likely to leave for a different federal position)
   c. overtime pay or compensatory time off
   d. special rates (e.g., higher rates of pay for an occupation or group of occupations nationwide, worldwide, or in a local area)
   e. critical position pay authority.

Changes to base compensation, such as a TLMS, are considered permanent changes to compensation. Targeted incentives, such as performance and retention incentives, are not permanent changes to compensation. The use of one-time incentives is less costly than permanent changes to income. Only when incentives are insufficient to achieve recruiting and retention goals should permanent income changes be tried.

Special rates and critical position pay authority reflect pre-CES efforts to support higher pay through permanent changes in compensation. Special rates reflect alternative pay schedules for an occupation or group of occupations that could be created when recruitment and retention efforts are insufficient to meet the government's manpower needs (OPM, 2013). An agency wanting a special salary rate must request and receive approval from OPM and coordinate its request with other agencies that have employees in the same occupational group and geographic area (OPM, 2013). These rates are capped at Executive Schedule Level IV. The outcome of this process could be pay schedules that mirror a TLMS applied to an entire occupation. Critical position pay authority is another compensation flexibility that can be requested by an agency head and approved by OPM. It fixes the rate of basic pay for a critical position above the current rate under the critical pay authority but not above Executive Schedule Level I.

Questionnaire Findings

Agencies varied substantially in the completeness of their responses to the employment and compensation questionnaire. The Office of the DoD CIO iterated with some agencies in an effort to improve the accuracy of the results. In this section, we provide an overview of the responses. Note that reporting agencies vary substantially in the size of their CES organization.

Twenty-seven questionnaires were returned across the six CES organizations. Sixty-three percent of questionnaire responses provided complete or partial feedback on employment questions as they pertain to cyber workers, 41 percent of questionnaire responses addressed a specific CES work role, and 67 percent of the questionnaire responses answered at least one of the compensation questions. Table 3.1 summarizes responses, focusing on vacancy rates (measured as the share of vacant positions relative to the number of vacant and filled positions) and loss rates (measured as the share of losses relative to reported employment).

Vacancy and loss rates in individual agency surveys varied substantially within a CES organization as well as across CES organizations. If we compare across CES organizations using the reported numbers, we find that the vacancy rate for cyber positions is typically less than the vacancy rate overall, with the exception of FLTCYBER. High vacancy rates can reflect high turnover or growing organizations. In Table 3.1, loss rates for cyber workers are typically lower than vacancy rates, suggesting that these organizations are likely growing. There is no clear evidence from Table 3.1 that these organizations are exhibiting higher turnover for cyber positions relative to their noncyber workforce.

Some of the organizations, such as FLTCYBER and ARCYBER, have vacancy rates in excess of 20 percent for cyber positions. High vacancy rates could support a pay increase if those positions remain persistently unfilled because offers are not accepted or postings do not attract qualified candidates.[3] ARCYBER did not provide information on job postings, offers, and acceptances in the past year, so we are unable to judge whether the vacancy rate reflects administrative issues or insufficient compensation. In one FLTCYBER agency survey, respondents indicated that 54 offers were made and 64 cyber positions were posted in the past year. Of the 54 offers, 39 offers were accepted, but only 20 had been hired. In another FLTCYBER agency, the number of job postings was not reported, but respondents stated that four offers were made, accepted, and hired. Another FLTCYBER agency reported 26 cyber job postings, 40 offers, 24 acceptances, and 23 hires. In only the first agency was there any indication that identifying qualified candidates and getting offers accepted and hired was a problem.

The only other CES organization whose reporting agencies provided information on job postings, offers, acceptances, and hires was AFCYBER. Similar to the FLTCYBER agency, AFCYBER agencies exhibited variation with no consistent evidence that vacancies, if posted, could not be filled. In general, questionnaire responses provide

---

[3] We focus on offers because candidates could apply. Making an offer is an indicator that the applicant met the agency's needs for the position. It is possible that offers differ relative to postings for reasons other than a labor shortage—for example, an agency is slow to review and interview qualified applicants.

Table 3.1
Summary of Key Employment Questions from CES Organizations

Organization (workforce size/questionnaires collected[a])	Overall	Cyber
USCYBERCOM (126/7)		
Reported workforce	31	unreported
Vacancy rate	16.2%	unreported
Loss rate	unreported	unreported
MARFORCYBER (395/1)		
Reported workforce	433	371
Vacancy rate	12.3%	11.7%
Loss rate	7.6%	7.5%
FLTCYBER[b] (1,050/6)		
Reported workforce	1,193	776
Vacancy rate	24.4%	28.5%
Loss rate	14.7%	9.1%
ARCYBER[c] (workforce size unknown/7)		
Reported workforce	5,469	5,136
Vacancy rate	20.8%	20.6%
Loss rate	27.4%	13.8%
AFCYBER (workforce size unknown/5)		
Reported workforce	115	32
Vacancy rate	26.3%	8.6%
Loss rate	16.5%	0.0%
JFHQ-DoDIN (113/1)	Insufficient information provided	

NOTES: Reported workforce sizes may be smaller than the true workforce if not all the appropriate human resources offices responded to the survey. Vacancy rates reflect the ratio of vacancies to vacancies and filled positions. Loss rate reflects the ratio of separations to the report workforce size.

[a] Workforce size is based on the number of people who organizations reported as eligible for CES conversion at the time the organization converted to the CES; this serves as a baseline for understanding whether the totality of reported workforce size reflects the overall workforce.

[b] FLTCYBER had multiple surveys completed by the same person.

[c] Consistent with how the DoD CIO presented the results, we assume that Joint Special Operations Command findings are included as part of a separately completed ARCYBER survey.

no or incomplete information on job postings, offers, acceptances, and hires for cyber positions broadly and for specific CES work roles. This information is critical to collect in order to demonstrate the existence of a labor shortage.

Moreover, most organizations did not provide employment and hiring information pertaining to specific CES work roles. Many CES organizations seemed unable to identify specific CES work roles as part of their responses. This highlights the need for positions to be continuously categorized according to CES work roles and for this information to be recorded so that it can be used to inform labor force analyses. In reviewing cases of CES work roles in which employment and hiring information was reported, there did exist some cases with high turnover. For example, at MARFORCYBER, the loss rate for Cyber Operators was 31 percent compared with an 8 percent loss rate of cyber positions in the rest of the organization. Another case was the authorizing official/designating representative work role in a FLTCYBER agency, where there was a loss rate of 60 percent, compared with an 11 percent loss rate for all DoD civilian cyber workers in that FLTCYBER agency. These examples are the exceptions. In most cases in which CES work roles were reported, there were too few associated positions to draw conclusions, no information provided, or incomplete information provided. There were also several cases identified in which information provided was inconsistent, potentially reflecting insufficient information (e.g., no positions were posted in the past year, but job offers were made). This information is critical to collect in order to demonstrate the need for a TLMS targeted at a specific work role.

As part of the compensation portion of the questionnaire, organizations were asked to report on whether specific recruitment incentives were used and, if they were used, the fraction of offers using the incentive and the average amount (if a cash incentive was used). At least one agency in USCYBERCOM, FLTCYBER, MARFORCYBER, and ARCYBER reported using cash incentives. Of those organizations reporting using this incentive, only FLTCYBER agencies reported using recruitment incentives in more than 5 percent of cases. Of the organizations reporting using the incentive, the reported values were in the range of $5,500 to $18,600, with most being in the range of

$9,000 to $14,000. Relocation incentives were used by at least one reporting agency in each organization (although one organization did not provide a response to the compensation portion of the questionnaire); USCYBERCOM, FLTCYBER, and ARCYBER used student loan repayments.

FLTCYBER, MARFORCYBER, and ARCYBER reported using cash retention incentives, but only MARFORCYBER provided additional information on coverage (2 percent) and average amount ($49,000). Overtime pay or compensatory time off existed in most reporting agencies. Use of special rates existed in at least one agency within FLTCYBER, ARCYBER, and AFCYBER. Critical position pay authority was used in at least one agency within FLTCYBER and AFCYBER. In all but two cases, no agency was able to report on the percentage of its cyber workforce who received or was eligible for overtime pay, special rates, or critical pay authority.

Our main takeaway from reviewing responses to the compensation portion of the questionnaire is that these agencies do not systematically track their use of recruitment and retention incentives. Only in the case of FLTCYBER do responses suggest regular use of these incentives for recruiting. Across all agencies, information on the use of these incentives for retention is incomplete.

Validation of Cash Incentive Use

Given the inconsistency of data reporting on recruitment and retention use, we collected data on cash incentives paid to DoD civilian cyber workers to validate what was suggested by responses to the employment and compensation questionnaire: Recruitment and retention incentives are not heavily used in the DoD cyber workforce. The Office of the DoD CIO provided a list of individuals in the DoD cyber workforce by CES work role (including CES and non-CES organizations) as of June 2019. Using this information, we extracted pay and personnel records for these individuals from September 2010 to September 2018, the last available time point as of the date of the analysis.[4] The pay records we used for this analysis identify award amounts paid during a specific pay period. To identify incentive awards, we use the DoD's Nature of Action codes in the civilian transaction files. The civilian transaction files contain records of personnel actions, such as position changes, change in duty station, and changes in employment status. Nature of Action codes include the issuance and payment of incentive awards (recruitment, relocation, retention, and performance). We identify three incentive categories, the values for which are set by OPM:[5]

- **performance award:** value is less than 5 percent of annual base and locality pay
- **recruitment/relocation award:** value is greater than or equal to 5 percent of annual base and locality pay but less than 25 percent of annual base and locality pay; payment occurs in first four years of employment
- **retention award:** value is greater than or equal to 5 percent of annual base and locality pay but less than 25 percent of annual base and locality pay; payment does not occur in first four years of employment.

The incentive transactions from the civilian transaction files contain dates and can be merged with a biweekly pay record in the civilian pay file. Table 3.2 summarizes the use of these incentives and confirms that recruitment, relocation, and retention awards are not heavily used in the DoD civilian cyber workforce.

The majority of the DoD cyber workforce received a one-time monetary incentive award in fiscal year (FY) 2018, but the awards

---

4 Pay and personnel records are provided by the Defense Manpower Data Center (DMDC). Pay records come from the civilian pay file, and personnel records are based on the civilian master file. Individuals entering the DoD cyber workforce after September 2018 were excluded from the analysis. Additional detail on the data is available in the section on DMDC data in Appendix A, and additional detail on the sample is available in the first section of Appendix B.

5 These definitions are based on our review of OPM guidelines and discussions with staff from the Office of the DoD CIO.

Table 3.2
FY 2018 Incentive Award Rates and Values

Award Category	Received Award	Mean	Standard Deviation	Median	Maximum
Performance award	69.24%	$1,332.78	$966.78	$1,102.08	$8,903.68
Recruitment/ relocation award	0.311%	$9,112.53	$4,441.81	$8,107.39	$25,003.34
Retention award	0.311%	$3,793.86	$2,728.85	$2,364.02	$11,265.56

NOTE: Incentive values are calculated conditional on receiving that award type as defined by Nature of Action codes and receiving a nonzero payment.

were primarily performance awards and not recruiting and retention awards. In reviewing earlier years for this workforce, we find similar rates of award receipt across these categories. To calculate the summary statistics in Table 3.2, we need to account for an incentive award date in the civilian transaction file not matching the exact pay period in which the award was paid out. We therefore look at four paychecks before and after the incentive award date and identify changes in an individual's gross pay amount that correspond to the above incentive payment definitions and include these in the calculation of the summary statistics.[6]

Summary and Discussion

Evidence exists that the U.S. cyber labor market is experiencing a shortage of cyber workers. Economic theory suggests this should lead to

---

6 In reviewing the timing of recruitment and retention awards recorded in the civilian transaction file relative to the payments in the civilian pay file, we determined that many of the recruitment and retention incentive awards were not reported in the award type and award amounts in the civilian pay file. However, these award values did seem to appear in gross pay, so we developed this approach, which enables us to recover the amount of the cash incentive awards that was consistent with the timing of the award reported in the civilian transaction file.

upward pressure on wages. However, evidence of a broader labor market shortage is not sufficient to justify an increase in CES pay because it does not demonstrate that existing compensation for CES positions is insufficient to meet the labor demanded by CES organizations. Existing compensation flexibilities, as well as nonmonetary benefits, might be sufficient to attract and retain quality cyber workers without the need to make permanent adjustments to pay. The Office of the DoD CIO needs to document cyber work role shortages in CES-covered agencies to provide a market-based reason for the use of permanent CES compensation flexibilities (notably the TLMS). Further, it needs to demonstrate that existing compensation flexibilities—specifically, recruitment, relocation, and retention incentives—are insufficient.

The Office of the DoD CIO sent an employment and compensation questionnaire to human resources offices in organizations that have converted or will convert to the CES in the near future. This questionnaire was intended to identify cyber labor shortages and document use of existing compensation flexibilities to address those shortages. Responses to the questionnaire varied in completeness. Inconsistencies in responses make it difficult to draw broad conclusions, particularly when responses represent a fraction of an organization's total civilian workforce (e.g., USCYBERCOM, AFCYBER) or when an organization's responses were largely incomplete (e.g., JFHQ-DoDIN). Consequently, this questionnaire produced insufficient information to recommend broad, immediate use of TLMS for the CES workforce. However, there are a few specific cases of CES work roles where evidence suggests high turnover and substantial vacancies indicative of a labor shortage (i.e., Cyber Operators in MARFORCYBER, authorizing official/designating representative in FLTCYBER). In these cases, an immediate wage adjustment through a TLMS would be warranted. We considered only market-based reasons for permanent pay adjustments; other reasons might exist for such adjustments, including compensation equity or mission risk.

Looking to the future, it is critical that CES organizations systematically collect and track the information requested in the questionnaire, particularly the items that we highlighted. This information is necessary to facilitate strong cases for a TLMS. Further, to facilitate labor force analyses, cyber positions and personnel need to be continuously categorized by their CES work role. In Chapter Six, we detail several recommendations toward this end. Additionally, given the inconsistencies in reporting on the questionnaires, we believe a process should be established that allows the Office of the DoD CIO to validate responses provided by the CES organizations. Data collection might be expanded to collect information on characteristics of offers made (e.g., incentives, starting step) and whether offers were accepted or rejected. This information could assist human resources managers in understanding what characteristics are associated with greater offer acceptance. In the next chapter, we characterize the DoD cyber workforce in greater detail and compare life-cycle earnings trajectories between DoD civilian cyber workers and private-sector cyber workers.

CHAPTER FOUR

Linking Select CES Work Roles to Private-Sector Occupations

The DoD CIO adopted the NIST NICE work roles for describing the jobs associated with the DoD civilian cyber workforce. As mentioned in Chapter One, the NICE work role descriptions delineate the KSATs required for these jobs. These NICE work roles are more precise than traditional occupations recorded in publicly available labor force survey data and were developed independent of the common occupation classification systems. Consequently, no direct comparisons exist between the NICE work roles and labor force survey data. To make these comparisons, we created a mapping between a DoD-identified subset of seven priority NICE work roles (referred to as *CES work roles*) and similar occupations reported in two publicly available labor force surveys that are nationally representative. We used these mappings to compare compensation of DoD civilian and private-sector workers in similar cyber jobs in Chapter Five, which might partly explain existing shortages in the supply of labor for certain DoD civilian cyber jobs (see "Theoretic Framework for Potential DoD CES Compensation Changes" in Chapter One).

Data Sources

We used several diverse sources of data for this project. Along with the NIST NICE Framework and government DMDC data on wages and government worker characteristics, we used data from the ACS and the Occupational Information Network (O*NET).

Data on private-sector workers and their wages come from the ACS, an annual survey administered by the U.S. Census Bureau.[1] These data provide information on numerous variables of interest, including geographic location, demographics, educational attainment, and earnings. The ACS identifies occupations using SOC taxonomy, a coding system used by federal agencies to classify workers into occupational categories. Up through 2017, the ACS used the 2010 SOC system. A new SOC classification system was released in 2018. We created a crosswalk to consistently code our occupations of interest in the ACS from 2012 to 2018.

O*NET is a database of KSATs necessary for a specific job based on surveys of analysts and industrial and organizational psychologists. In addition to identifying needed KSATs for an occupation, O*NET maps KSATs to SOC codes. As described in detail below, we use O*NET to create a correspondence between the NICE government work roles and the SOC codes found in the ACS. Additional details about these data sets can be found in Appendix A.

Methodology

The DoD CIO identified seven critical CES work roles for our labor market analysis. We develop a mapping between these CES work roles and traditional occupations by comparing the detailed lists of unique tasks associated with each of the 52 CES work roles and the corresponding KSAs with occupational classifications used in nationally representative labor force surveys. We define occupations according to

---

1 The Occupational Employment Statistics (OES) data also contain information on employee wages. Unlike the ACS, which gathers data directly from employees, the OES gathers data from employers. However, the OES does not have information on such worker characteristics as gender, educational attainment, and age. Because the government workforce is different than the private-sector workforce on these characteristics, we used the ACS data instead of the OES data to allow us to better compare private-sector wages with government wages.

the 2018 SOC.[2] For occupational KSATs, we used information developed by the O*NET based on surveys of workers in all occupations.[3]

The NIST NICE Cybersecurity Workforce Framework groups cyber-related jobs into seven overarching categories based on common cybersecurity functions. From these seven categories, the framework identifies 33 distinct areas of cybersecurity work. The 52 cyber work roles are the most detailed groupings of cybersecurity work, each of which is composed of specific KSAs that are required to perform the unique tasks associated with each work role. Many of the KSAs are common across many of the work roles and reflect baseline skills and common knowledge that are widely applicable to most cybersecurity jobs—both public and private. The tasks, however, are not only unique to each of the work roles listed in the NICE framework but might also be distinct from the cybersecurity tasks performed in the private sector. In other words, while many KSAs might be the same in the public and private sectors, the tasks are not.[4]

First, we developed a text clustering algorithm to compare KSATs in the seven CES NICE work roles and in O*NET and narrow the list of possible occupational code matches based on text similarities in job roles and task descriptions (see Appendix B for additional details). We focused on O*NET jobs whose categorization began with 11 (management), 13 (business and financial operations), 15 (computer and mathematical), and 33 (protective service) because these were the most similar to the jobs that we were matching to. The algorithm then provided our team an efficient and repeatable method to eliminate all of the O*NET occupation codes within these categories that shared no tex-

---

2 We initially used 2010 SOC occupation codes. We mapped 2010 SOC occupational codes to 2018 occupational codes. See BLS, 2020c, and O*NET OnLine, undated b.

3 We link O*NET occupational information to 2018 SOC occupations using a linkage provided by O*NET.

4 The fact that government and private-sector jobs differ more on tasks performed than they do on KSAs needed and the fact that some of these tasks are specific to government work and not found in the private sector make it difficult to exactly match government work roles to private-sector jobs, and this is a limitation of our work. As noted below, several government work roles are mapped to the same SOC, even though these work roles may have distinct wages that reflect their different tasks.

tual similarities to the seven CES work roles and reduced the number of possible occupation code matches from 192 to between 20 and 30 for each CES work role of interest.

Using this reduced set of potential matches for a given specific NICE work role, we asked RAND researchers with subject-matter expertise in cyber issues to review the remaining potential matches. They were asked to identify the best possible O*NET job code matches for each CES work role based on work role descriptions, KSATs, and their subject-matter expert (SME) understanding of these work roles. We asked RAND cyber SMEs to review potential matches for each of the seven CES work roles.[5]

We asked each RAND SME to assign a numeric value for every O*NET occupation code in our reduced set that could be matched to a CES NICE work role to indicate the degree of similarity between the O*NET occupation and the NICE work role of interest. We also asked each SME to identify a best match and, if necessary, a second- and third-best match. For example, an SME would review the title, work role description, and associated KSATs for each cyber work role and compare that with the O*NET code occupation descriptions, tasks, and associated skills. The SMEs also used their expertise on the responsibilities and functions of the CES work roles and the O*NET occupation codes when making their comparisons. The SMEs were also asked to list any critical tasks and skills that were essential to each CES NICE work role to ensure that the most appropriate O*NET job code had been matched to the CES NICE work role and to explain their thought processes and reasons for their rankings.

To establish a common ranking, we compared the numeric value assigned with each O*NET occupation code by each SME. Where each of the SMEs provided a common ranking, we adopted them. Where SMEs identified different rankings, we considered the additional information provided on critical skills and tasks for each CES work role and SMEs' rationale for their rankings. The end result of the

---

5 This process was done twice. The first time, we used five researchers. Given the consistency of our findings, subsequent reviews used feedback from only three researchers.

RAND cyber SME review was a ranking of O*NET job codes that best fit each of the seven CES work roles.[6]

Linkages and Analysis

The linkages identified by the RAND SMEs between the CES work roles and SOC occupational titles are reported in Table 4.1, using the methodology described in the previous section. We refer to 2018 SOC occupational titles as *ACS occupational titles* to distinguish that the ACS uses only the first six digits of the 2018 SOC occupational titles. The first ACS occupational title corresponding to each CES NICE work role is considered the best possible match. If other ACS occupation titles were also identified as appropriate matches, the alternates are listed in descending rank order. Finally, we also include the OPM cybersecurity codes that correspond to NICE work role and NICE job codes. The federal government uses these OPM codes to identify positions that require the performance of information technology, cybersecurity, and other cyber-related functions. These three-digit OPM cybersecurity codes map directly to the work roles described in the NICE Framework and will be used to refer to the NICE work roles throughout the report.[7]

---

[6] The ACS uses six-digit SOC occupation codes instead of the more detailed eight-digit O*NET job codes. As a result, we conducted an additional comparison between the job titles and descriptions to ensure consistency of our best matches across the data sets. Additionally, the ACS updated its occupational codes between 2017 and 2018 to reflect a classification change from the 2010 SOC to the 2018 SOC. To address changes in occupational coding, we compared our seven cyber work roles and the corresponding O*NET job codes (which used the 2010 SOC) with the 2018 SOC. Because some of the ACS job codes were changed and new job titles were introduced between the 2010 SOC and the 2018 SOC (e.g., the job title "Software Quality Assurance Analysts and Testers" was introduced only in the 2018 ACS), we both adjusted for the change in job codes and elected not to include new 2018 job codes in order to maintain a consistent sample of job codes over time. Overall, the differences between the 2017 and 2018 ACS occupational classifications were minimal and easily accounted for in our analysis.

[7] For a crosswalk of NICE work role identifications to OPM cybersecurity codes, see Newhouse et al., 2017.

Table 4.1
Seven NICE CES Work Roles Linked to SOC Occupational Titles

NICE CES Work Role	NICE Job Code	2018 ACS Occupational Title[a]	OPM Cybersecurity Code	2018 ACS Occupational Code
Cyber Defense Analyst	PD-CDA-001	Information Security Analyst	511	15-1212
Cyber Operator	CO-OPS-001	Computer Network Architect	321	15-1241
		Computer Systems Analyst[b]		15-1211
		Computer Programmer[b]		15-1251
Security Control Assessor	SP-RSK-002	Information Security Analyst	612	15-1212
		Network and Computer Systems Administrators		15-1244
Software Developer	SP-DEV-001	Software Developers	621	15-1252
		Computer Programmer		15-1251
Systems Security Analyst	OM-ANA-001	Information Security Analyst	461	15-1212
		Computer Systems Analyst		15-1211
Authorizing Official	SP-RSK-001	Computer and Information Systems Manager	611	11-3021
Cyber Defense Incident Responder	PR-CIR-001	Information Security Analyst	531	15-1212

NOTE: ACS occupational code is the first six digits of the 2018 SOC occupational code.
[a] 2018 ACS Occupational Titles are listed in order of best possible match. At the time of writing, the 2018 ACS Occupational Titles and Codes were the most current. In the 2020 ACS Occupational Titles and Occupational Codes, Software Developers (15-1252) and Software Quality Assurance Analysts and Testers (15-1253), which were two separate codes in the 2018 ACS Occupational Titles and Codes, were combined into one new code, Software Developers & Software Quality Assurance Analysts and Testers (15-1256). Among the ACS Occupational Titles and Codes included in our analysis, this was the only update between 2018 and 2020 and it does not impact our analysis or findings.
[b] Indicates equal ranking.

In three cases, such as in the case of the CES NICE work role Cyber Defense Analyst, SME respondents unanimously agreed that one private-sector job—in this case, Information Security Analyst—was the most appropriate match. SMEs also generally agreed on the essential KSATs that would be needed for each of these jobs, such as the inherently defensive nature of the two jobs and the emphasis on monitoring, continuing assessment, reporting, and timely reaction. The SMEs also cited the need to understand how to use relevant cyber defense and information security tools, particularly the automated ones that often function as the foundation of computer and network defense.

In the remaining four cases, the SMEs matched two or three possible occupational titles to the identified CES work roles.[8] For example, the SMEs identified (1) Computer Network Architect, (2) Computer Systems Analyst, and (3) Computer Programmer as the best matches to the CES NICE Cyber Operator work role. Respondents agreed that the Computer Network Architect job was the best match overall because it focused on the development of computer and systems architecture in addition to the other functions of a Cyber Operator, such as troubleshooting, developing, gaining and maintaining access, and understanding how to attack a system. They also agreed that the Computer Systems Analyst and Computer Programmer occupations were equally appropriate, based on such shared occupational characteristics as the ability to design a system, familiarity with troubleshooting and automation, and developing new techniques for gaining and keeping remote access.

---

8 For cases with multiple SOC matches per CES work role, one approach would be to devise weights based on the tasks in common between each private-sector job and the government work role and then create a theoretical private-sector job that better approximates the government job of interest. However, we decided to use only the best match in our analyses because this was the most straightforward approach and avoided introducing additional subjectivity around the exact weighting of various private-sector jobs.

Discussion and Findings

This linkage exercise identified several important distinctions between the relevant NICE CES work roles and private-sector cyber occupations. The first distinction relates to the offensive cyber activities performed in some of the NICE CES work roles. Offensive cyber operations and activities are an inherently governmental or military cyber job and therefore will not have a direct match in the private sector. In fact, several SMEs noted that many of the tasks and activities conducted by individuals in more operationally offensive cyber work roles, such as the Exploitation Analyst work role, not only lack a direct match to the private sector but would also be illegal to conduct. As noted above, although the KSAs required for an Authorizing Official or a Cyber Operator might be similar to the KSAs required to perform certain cybersecurity jobs in the private sector, the tasks are distinct. Therefore, although the NICE work roles were created to identify cyber roles regardless of who may be conducting the activity (government or private sector), this example demonstrates that there are some tasks and responsibilities of select NICE work roles that would be conducted by individuals in the U.S. military or that would involve taking actions on behalf of a government agency. Therefore, aligning some of the NICE CES work roles with the private-sector cyber jobs might not always provide an ideal linkage, given the unique differences in the tasks that these jobs require the individual to perform.

Setting this difference aside, the SMEs agreed that many of the KSAs necessary to conduct offensive computer network exploitation activities still translate to other nonoffensive roles in the private sector, such as Computer Network Architects, Computer Systems Analysts, and Computer Programmers. Upon first glance, these private-sector jobs might not seem to be good fits for their CES NICE work role counterparts because the titles appear unrelated and the summarized descriptions of these jobs are quite different. However, understanding the listed KSAs of these jobs, as well as having a familiarity with how these jobs are performed and what they accomplish in the real world, allowed us to identify matches between CES work roles (as public-sector jobs) and private-sector jobs that might otherwise have been overlooked. Our approach of using available job data and leveraging cyber SMEs with real-world knowledge makes this crosswalk both methodologically rigorous and valuable to our project.

Another important distinction between the seven CES work roles and private-sector cyber occupations is the different interpretation of security. For the CES work roles, particularly the Security Control Assessor work role, security is defined as cyber-related, which does not include physical security. Most private-sector jobs identified in our O*NET and subsequent ACS search that included *security* in the title also included functions of both cyber (or virtual) and physical security. When using the list of O*NET job titles, the SMEs identified the private-sector job of Security Management Specialist as the best overall match for the CES NICE work role Security Control Assessor. However, the SMEs gave the match a lower score because the physical security tasks of the private-sector job do not match the cybersecurity-focused tasks of what one would do as part of the CES NICE work role. Ultimately, this issue became irrelevant once we reconciled the matches to the 2010 SOC and 2018 SOC occupation codes. In the 2018 SOC occupation codes, the Security Control Assessor job that SMEs identified as a close (but not perfect) match was incorporated into two existing positions, Information Security Analysts and Network and Computer Systems Administrators, both of which share a similar interpretation of *security* as defined in the cyber work role descriptions. Therefore, the SMEs' concern regarding the physical security aspects of the outdated Security Control Assessor position were resolved.

Our SMEs also identified a similar distinction between the CES NICE work role and private-sector uses of management functions. This was most apparent when comparing the senior level authorizing and management functions of the CES NICE work role Authorizing Official with possible ACS jobs that also included management and authorizing functions. Although our SMEs concluded that there was no direct match for the Authorizing Official work role, the best ACS match and the job that required many of the same KSAs and associated tasks was the Computer and Information Systems Manager job title.

This exercise also provided some interesting insights. First, the job Information Security Analyst was identified as the best match for four of the seven total CES work roles. In addition, the job Computer Systems Analyst was listed as either the best match or second-best match for three of the seven total CES work roles. These findings highlight the broad utility of the KSAs of Information Security Analysts and Computer Systems Analysts for the CES cyber work force.

CHAPTER FIVE

DoD Civilian Cyber Worker and Private-Sector Cyber Worker Comparisons

When a firm experiences a labor shortage that reflects pay insufficiency (as opposed to administrative difficulties in job posting and hiring), a comparison of the firm's workforce to workers at similar firms can characterize potential reasons for that insufficiency and provide a reference point for adjusting compensation. Data provided to us on labor demand did not point to a universal labor shortage for DoD civilian cyber workers or for specific CES organizations. This might reflect the quality of the data provided, or it might indicate that the perceived shortage of cyber workers that motivated the creation of the CES is more nuanced.

In this chapter, we use administrative and survey data to compare DoD civilian cyber workers and private-sector cyber workers' characteristics and pay. We do not find overall differences in pay between DoD and private-sector cyber workers, but we do find differences in some worker characteristics and pay over the life cycle and in certain local areas. For example, broad adjustments to provide higher pay in certain areas for all federal workers (typically reflecting cost of living differences) do not eliminate these differences for cyber workers.[1]

We first compare cyber worker characteristics, such as gender, age, and educational attainment. Then we present predicted life-cycle earnings (i.e., pay trajectories) that reflect only pay differences between private-sector and DoD civilian cyber workers by controlling for worker characteristics, location, and work role. These reference points provide

[1] In 1990, Congress passed the Federal Employee Pay Comparability Act, which allowed for additional pay for federal employees based on the geographic designations where they worked. This was an effort to help close the wage gap between the federal and private sectors.

important labor market benchmarks that can support targeted CES pay adjustments when labor shortages exist.

DoD and Private-Sector Data on Employment and Pay

For information on DoD civilian worker characteristics and pay, we use administrative data from the DMDC Civilian Master File (CMF), which provides information on demographic characteristics, and the DMDC Civilian Pay File (CPF), which provides information on pay and location. Neither of these files categorize DoD civilian cyber workers by work role. The Office of the DoD CIO provided a list of individuals in the DoD cyber workforce by CES work role (including CES and non-CES organizations) as of June 2019. We identify personnel and pay records for that workforce in our most recent DMDC civilian data (September 2018) and use that as the basis for our sample of DoD civilian cyber workers.[2] We drop individuals who had different occupations listed in the September 2018 and June 2019 files.[3] We assume that the remaining individuals were in the same cyber work role in June 2019 and nine months prior (September 2018). Because that assumption becomes more tenuous as we go further back in time, we restrict our DMDC sample to just September 2018. Other DoD civilian cyber workers in cyber work roles of interest in September 2018 likely had changed to noncyber occupations or left the DoD civilian work force

2 Although we have DMDC data through June 2019, the ACS data for 2019 were not expected to be available until fall 2020. We use end-of-FY September 2018 DMDC data as our complement to the ACS 2018 data.

3 The Office of the DoD CIO provided a list of 65,440 nonduplicate individuals in the DoD cyber workforce in June 2019; 64,726 of these individuals appear in the June 2019 CMF, and 61,760 of those individuals also appear in the September 2018 CMF (the difference largely reflects new hires between September 2018 and June 2019). We exclude noncyber work roles, which further reduces the sample to 52,902 individuals. Merging these records with the September 2018 CPF and eliminating individuals who were in our work roles of interest but had different occupations listed in the September 2018 and June 2019 files reduces the sample to 52,744 individuals. Finally, restricting to individuals with a September 29, 2018, paycheck and dropping workers with fewer than 70 regular hours over a two-week period, we are left with 50,968 individuals. These are the observations that go into the pay regressions.

by June 2019. but they are missing from our analysis; we have no way of identifying these individuals or ability to estimate their number.[4]

For information on private-sector wages and worker characteristics, we use survey data from the ACS Public Use Microdata Sample (PUMS). The ACS is conducted on an ongoing basis and updated yearly. These data contain such individual characteristics as gender, education, occupation, geographic location, and pay.[5] As a measure of pay, we use wage or salary income for the past 12 months.[6] Although the ACS samples 3.5 million households every year, sample sizes can be too small to produce accurate estimates for a specific occupation in a specific geographic location (e.g., cyber workers in the Washington, D.C., metro area). To address this limitation, we pooled ACS data from 2012 to 2018. Pay measures are adjusted for inflation and reported in 2019 dollars. We identify private-sector workers in jobs that are most comparable to the seven cyber work roles using the linkages from Chapter Four.[7]

---

4 We spent a substantial amount of time trying to use the characteristics in the CMF (with interactions between occupation, Unit Identification Code, grade, pay plan, education, and year) to predict the probability that an individual would be part of the CES in 2018. This would have allowed us to find individuals who looked like CES members before the CES was created and to trace their career trajectory through time both before and after their organizations converted to the CES. It also would have allowed us to identify people who were likely in the CES but who left between September 2018 and June 2019. Unfortunately, with the variables available, our multinomial logit specification did not converge, and we were not able to come up with a reliable, repeatable algorithm to predict potential CES status with confidence, given the available explanatory variables available.

5 An alternative data source is the OES. OES, as a survey of employers, has detailed information about income by occupation but no information on demographic characteristics. Because demographic characteristics vary across workers in cyber and other occupations, and across the DoD civilian and private sectors, we chose ACS as our preferred data source. In addititon, the Census Bureau maintains individual-level administrative wage data for private-sector firms that also provide information on worker characteristics, but this resource is not publicly available. For additional details on the OES and ACS, see Appendix A.

6 We chose this measure because it is the most similar to basic pay for DoD workers (including base and locality pay). There are several measures of income in the ACS. Besides wage or salary income for the past 12 months, there are total earnings, including from business and farm income, and total pre-tax income or losses from all sources.

7 The final unweighted ACS PUMS of cyber workers used in our analysis had the following number of observations by year: 2012: 29,606; 2013: 32,441; 2014: 33,931; 2015: 35,633; 2016: 37,821; 2017: 39,836; 2018: 40,662.

We aim to make the ACS PUMS data used in our analysis as similar as possible to the DMDC data in terms of worker characteristics and geography, but we note that the two data sources differ in two important ways. First, the ACS is a survey and the DMDC data are administrative records. Administrative records are typically considered more accurate measures than survey responses because the latter are subject to recall problems and response bias. Second, the ACS is a repeated cross-section, meaning that different people appear in the survey each year, while DMDC data are a panel data set, tracking the same individuals for the duration they work for the DoD. Still, by using only September 2018 DMDC data for our analysis, we are essentially using a cross-section of the DMDC data. For additional information on the ACS and DMDC data, see Appendix A.

In the next section, we compare demographic characteristics across the DoD civilian and private-sector cyber workforces. These comparisons lay the groundwork for why demographic characteristics need to be controlled for in estimating pay trajectories.

Comparison of Cyber Worker Characteristics

In 2018, the cyber workforce constituted 8 percent of the DoD civilian workforce and 3 percent of the private-sector workforce.[8] We compare these two workforces across a variety of dimensions, including gender, age, citizenship, veteran status, and weekly work hours (see Table 5.1).

The percentage of male workers is effectively the same across the two workforces. Men constitute 78 percent of the DoD civilian cyber workforce and 76 percent of the private-sector cyber workforce. The DoD civilian cyber workforce is older on average, by approximately seven years at the mean (48 versus 41) and ten years at the median (50 versus 40). All DoD civilian cyber workers are U.S. citizens; the U.S.

---

[8] In rough raw numbers, just under 51,000 cyber workers are in our total 2018 DMDC sample of 660,000, and just over 4 million cyber workers are in our total 2018 OES sample of 145 million. We use OES data to calculate workforce size because OES coverage of workers far exceeds that of the ACS. We use ACS data for the remainder of the analysis—descriptive statistics and regression modeling—because they have richer information on worker characteristics.

Table 5.1
Select Demographic Characteristics: DoD Civilian and Private-Sector Cyber Workforces

Demographic	DoD Civilian Cyber Workforce	Private-Sector Cyber Workforce
Male	79.0%	75.9%
Mean (median) age, in years	48.2 (49.8)	41.1 (40.0)
U.S. citizen	100.0%	85.5%
Veteran	46.0%	6.4%
Weekly work hours (mean)	40.0	43.0

NOTE: All differences between the DoD civilian cyber workforce and the private-sector cyber workforce are statistically significant at the 1 percent level.

citizen share of the private-sector cyber workforce is lower (86 percent). DoD civilian cyber workers are more than seven times more likely than private-sector cyber workers to be veterans (45 percent versus 6 percent). Finally, private-sector cyber workers work slightly more hours per week: an average of 43, versus 40 for DoD civilian cyber workers.[9]

In Figure 5.1, we compare educational attainment among DoD civilian and private-sector cyber workers. Compared with private-sector cyber workers, DoD civilian cyber workers are

- more likely to have less than a college degree (38.6 versus 29 percent), with the majority of that difference explained by the considerably higher share of DoD civilian cyber workers who are high school graduates (21.9 versus 5.7 percent)
- less likely to have a college degree or more (61.4 versus 70.5 percent), with the majority of that difference explained by the considerably lower share with a college degree (36.6 versus 46.5 percent); in contrast, DoD civilian cyber workers are slightly *more* likely to have a master's degree or more (24.8 versus 24.0 percent).

9 We note, as discussed earlier in this section, that the small difference in work hours could reflect the fact that we are using two different types of data sources, administrative data (DMDC) and survey data (ACS).

Figure 5.1
Educational Attainment of DoD Civilian and Private-Sector Cyber Workforces

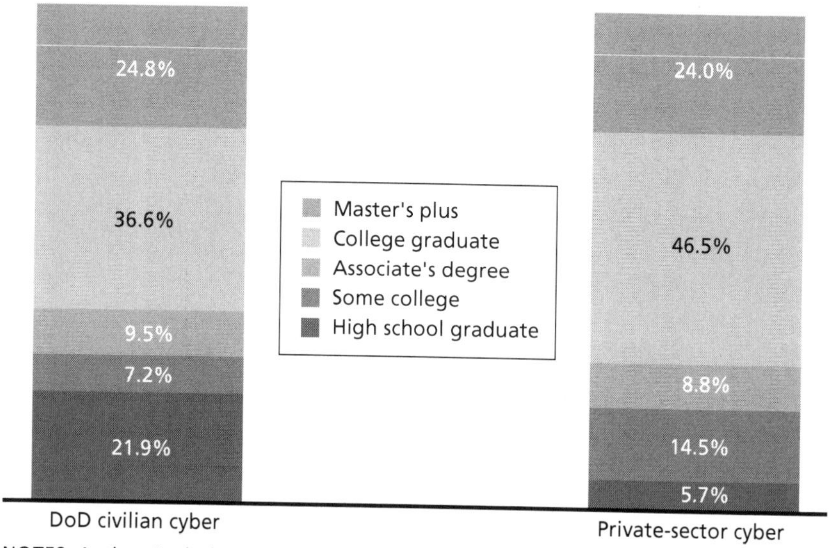

NOTES: Authors' tabulations using September 2018 CMF from the DMDC matched to persons identified and categorized as part of the DoD cyber workforce by the Office of the DoD CIO in June 2019 and the 2018 ACS matched to cyber work roles using a crosswalk between DoD cyber work roles and private-sector occupations. Numbers in the private-sector column do not add to 100 percent because persons without high school degrees are not shown.

In Figure 5.2, we compare average years of potential experience by cyber work role.[10] Civilian cyber workers have more years of potential experience, on average, across all work roles.[11] The differences in average years of potential experience are two years or greater, with the largest differences appearing in the following work roles:

- Authorizing Official/Designating Representative (611), with a difference of 5.8 years

---

[10] We use *years of potential experience*—defined by age and educational attainment, as described in Appendix B—in lieu of years of actual work experience because we do not observe years of work experience outside the DoD in DMDC data and the ACS does not contain information on years of experience.

[11] We observe the same pattern across the entire DoD civilian workforce.

Figure 5.2
Average Years of Potential Experience of DoD Civilian and Private-Sector Cyber Workforces

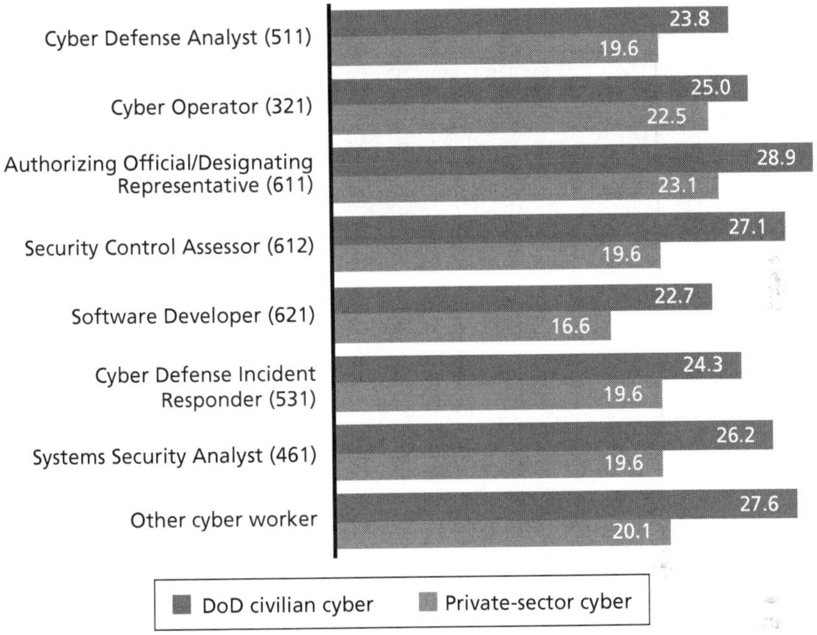

NOTES: Authors' tabulations using September 2018 CMF from the DMDC matched to persons identified and categorized as part of the DoD cyber workforce by the Office of the DoD CIO in June 2019 and the 2018 ACS matched to cyber work roles using a crosswalk between DoD cyber work roles and private-sector occupations. Recall that the Cyber Defense Analyst (511), Security Control Assessor (612), Cyber Defense Incident Responder (531), and Systems Security Analyst (461) work roles match to the same ACS occupation, so the values shown here for private-sector workers in these work roles are equivalent.

- Security Control Assessor (612), with a difference of 7.5 years
- Software Developer (621), with a difference of 6.1 years
- Systems Security Analyst (461), with a difference of 6.6 years
- Other cyber work roles, with a difference of 7.5 years.

Given how years of potential experience are constructed, the results shown in Figure 5.2 are consistent with the fact that the DoD civilian cyber workers are older on average and are more likely to have less than a college degree.

Finally, we compare the geographic distribution of the DoD civilian and private-sector workforces. We present a simplified view in Table 5.2, focusing on the three locality pay areas (LPAs) that constitute the largest shares of the DoD civilian cyber workforce: Washington-Baltimore-Arlington, DC-MD-VA-WV-PA; San Diego Carlsbad, California; and Los Angeles-Long Beach, California, which constitute 20.0 percent, 4.5 percent, and 2.9 percent of the DoD civilian cyber workforce, respectively. Collectively, these three LPAs constitute 27.4 percent of the DoD civilian cyber workforce.[12] The rest of the DoD civilian sector workforce is not as concentrated in these three LPAs, which constitute just 20.1 percent of the broader civilian workforce. The same is true for the private-sector workforce, where these three LPAs constitute just 12 percent and 9.7 percent of the cyber workforce and the rest of the workforce, respectively.

To summarize, compared to private-sector cyber workers, DoD civilian cyber workers are older and less likely to have a college degree or more and therefore have more years of potential experience, on average. Also, DoD civilian cyber workers are more likely to be U.S. citi-

Table 5.2
Geographic Distribution of DoD Civilian and Private-Sector Cyber Workforces

	DoD Civilian Sector		Private Sector	
Locality	Cyber Workforce	Rest of the Workforce	Cyber Workforce	Rest of the Workforce
Washington-Baltimore-Arlington, DC-MD-VA-WV-PA	20.0%	14.1%	6.2%	3.1%
San Diego Carlsbad, CA	4.5%	3.2%	1.0%	0.9%
Los Angeles-Long Beach, CA	2.9%	2.8%	4.8%	5.7%

12 We note that Augusta, Georgia (where ARCYBER is located), and San Antonio, Texas (where AFCYBER is located), are coded as Rest of the United States in the DMDC data and therefore cannot be examined separately.

zens and to be veterans than private-sector cyber workers are. We find only small differences in the percentage of male workers and the average weekly work hours across the two workforces. Finally, we find that the private-sector cyber workforce is considerably less likely to be concentrated in the three LPAs that constitute the largest share of the DoD civilian cyber workforce.

Comparison of Cyber Worker Pay

As we demonstrate, the characteristics of DoD civilian cyber workers and private-sector workers differ in important ways, and these differences should be accounted for in pay comparisons. For example, pay generally rises with age, reflecting an experience premium. Since DoD civilian cyber workers are much older on average than private-sector cyber workers, if we observed the same pay on average for DoD and private-sector workers, it may indicate lower life-cycle pay for DoD civilian cyber workers.

Using our ACS and DMDC samples, we estimate life-cycle pay trajectories—average pay by year of potential experience—separately for private-sector and DoD civilian cyber workers.[13] We begin by estimating pay regressions—one for the ACS sample and one for the DMDC sample—controlling for cyber work role, gender, local pay area, education category (e.g., less than a bachelor's degree, bachelor's degree, and master's degree or more), and years of potential experience in five-year splines.[14] We use the estimates from these regressions to construct average pay trajectories in a given LPA and for a given work

---

13 Several researchers have documented that individuals may underreport wages on surveys, such as the ACS (see, for example, Moore, Stinson, and Welniak, 2000). Because wages in the ACS are self-reported, underreporting is a potential risk, which might bias private-sector earnings trajectories lower, thus reducing differences between DoD civilians and private-sector civilians. We explore potential underreporting in the ACS in Appendix A.

14 For the private-sector regressions, in which we include multiple years of data in the sample, we also include year dummies in the regressions.

role and education level, comparing these wage trajectories across the DoD and private sectors.[15]

Our model allows for different pay *levels* by work role but not different pay *trajectories* by work role. In other words, our methodology will not reveal whether average wages for one cyber work role have a different trajectory over the course of a career than another. We impose this constraint because the small size of most of the cyber work roles makes it difficult to identify independent pay trajectories by work role and location. Additional details on the construction of the sample and the pay regressions can be found in Appendix B. Here, we discuss wage trajectories for workers with a bachelor's degree in the Washington-Baltimore-Arlington, DC-MD-VA-WV-PA LPA, which we refer to as the *Washington, D.C. LPA*. At the Office of the DoD CIO's request, we produced results for 12 additional LPAs.[16] Appendix C contains the detailed pay trajectory data for the Washington, D.C. LPA and the others LPAs.

Figure 5.3 shows the results for Cyber Defense Analyst (511) among workers with a bachelor's degree in the Washington, D.C., area. We observe that Cyber Defense Analyst (511) has a similar DoD civilian pay trajectory *and* matches to the same ACS occupation code as three other cyber work roles: Security Control Assessor (612), Systems Security Analyst (461), and Cyber Defense Incident Responder (531). Therefore, Figure 5.3 represents the findings for all four of these cyber work roles. As the figure shows, there is a DoD civilian pay premium of roughly $14,000 at hiring, but that shrinks as years of potential experience increase. At 19 years of potential experience, the two trajectories cross; beyond that point, a private-sector pay premium emerges. The private-sector pay premium is relatively small, roughly $3,000, and remains small through 30 years of potential experience. After 14 years of potential experience, the difference is not statistically significant.

---

15 In all predictions, we assume that the prediction reflects the coefficient associated with men. Although differences in pay exist between men and women, we do not reflect these in our comparisons.

16 The additional 12 LPAs are: Hawaii, Huntsville, Indianapolis, Los Angeles, New York, Philadelphia, Sacramento, San Diego, San Francisco, Seattle, St. Louis, and Tucson.

Figure 5.3
Comparison of Predicted DoD Civilian and Private-Sector Pay Trajectories for Cyber Defense Analysts in the Washington, D.C. LPA

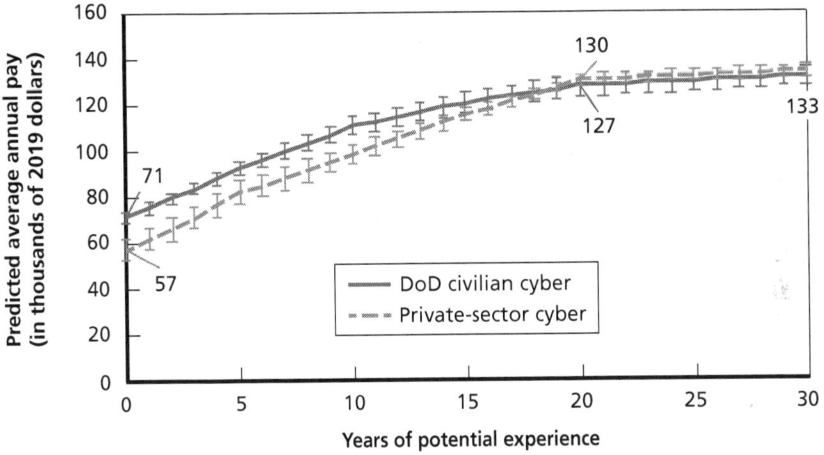

NOTES: Predictions are generated by regression models that account for gender, LPA, educational attainment, and years of potential experience using 2012–2018 ACS and September 2018 DMDC data. Predictions are for men with bachelor's degrees in the Washington, D.C. LPA. Error bars correspond to the prediction's 95 percent confidence interval.

Figure 5.4 shows results for Authorizing Official/Designating Representative (611). The pattern is similar to what we saw in Figure 5.3, however, the DoD pay premium is larger, approximately $17,000, at hiring. That premium steadily shrinks as years of potential experience increase until 20 years of potential experience, where the two trajectories converge.

Figure 5.5 shows results for Software Developer (621). There is a DoD civilian pay premium of roughly $10,000 at hiring, but that shrinks as years of potential experience increase. At 14 years of potential experience, the two trajectories cross; beyond that point, a private-sector pay premium emerges. The private-sector pay premium reaches a maximum of roughly $12,000 at 20 years of potential experience and remains at roughly that amount through 30 years of potential experience.

Figure 5.4
Comparison of Predicted DoD Civilian and Private-Sector Pay Trajectories for Authorizing Official/Designating Representative in the Washington, D.C. LPA

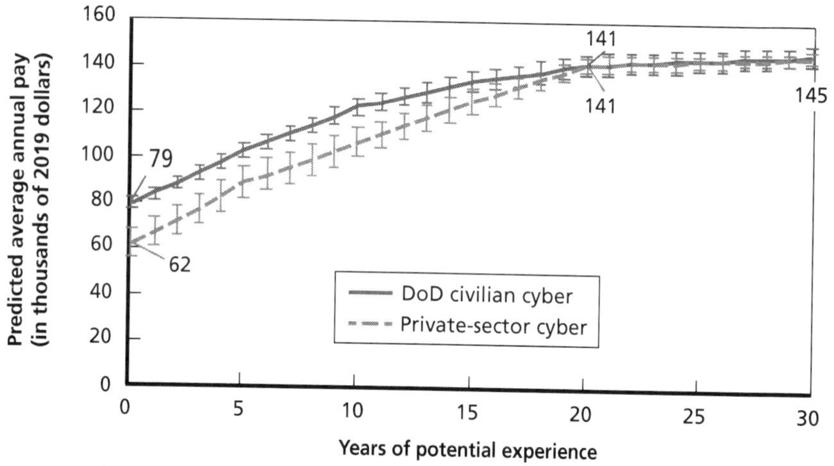

NOTES: Predictions are generated by regression models that account for gender, LPA, educational attainment, and years of potential experience using 2012–2018 ACS and September 2018 DMDC data. Predictions are for men with bachelor's degrees in the Washington, D.C. LPA. Error bars correspond to the prediction's 95 percent confidence interval.

Finally, Figure 5.6 shows results for Cyber Operator (321). There is a small DoD pay premium at hiring (approximately $6,000). At five years of potential experience, the trajectories overlap until reaching ten years of potential experience, when the private-sector trajectory pulls away from the DoD civilian trajectory. The private-sector pay premium reaches a maximum of $20,000 at 20 years of potential experience and persists at that amount through 30 years of potential experience, a difference that is statistically significant.

Key Takeaways

There are some similarities and some differences in the pay trajectory results across the cyber work roles in the Washington, D.C., area. In

Figure 5.5
Comparison of Predicted DoD Civilian and Private-Sector Pay Trajectories for Software Developer in the Washington, D.C. LPA

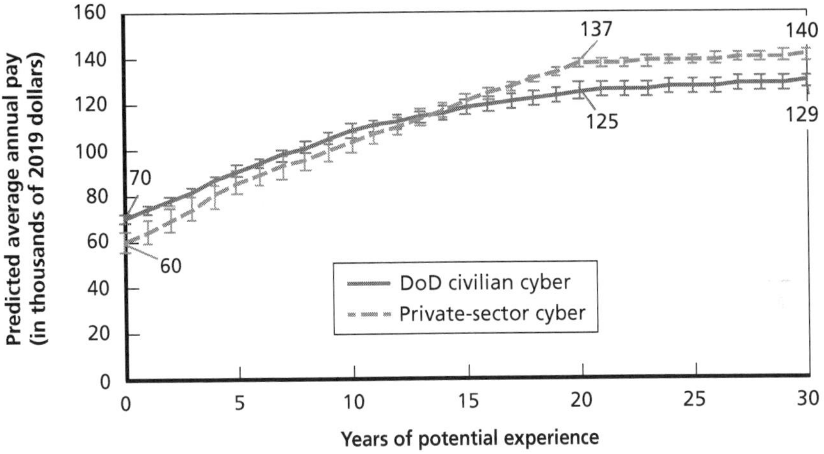

NOTES: Predictions are generated by regression models that account for gender, LPA, educational attainment, and years of potential experience using 2012–2018 ACS and September 2018 DMDC data. Predictions are for men with bachelor's degrees in the Washington, D.C. LPA. Error bars correspond to the prediction's 95 percent confidence interval.

all cases, there is a DoD civilian pay premium at hiring. It is largest for Authorizing Official/Designating Representative (611)—roughly $17,000—and smallest for Cyber Operator (321)—roughly $6,000. However, that DoD civilian pay premium shrinks as years of potential experience increase, at which point a private-sector pay premium emerges for all but one work role—Authorizing Official/Designating Representative (611). The DoD civilian pay premium vanishes earliest in the career for Cyber Operator (321)—at ten years of potential experience. The private-sector pay premium that emerges midcareer for Cyber Defense Analyst (511)—and the three other cyber work roles that have similar DMDC wage trajectories and identical ACS wage trajectories, Security Control Assessor (612), Systems Security Analyst (461), and Cyber Defense Incident Responder (531)—is small (less than $3,000) and remains small after 20 years of potential experience. In contrast, the private-sector pay premiums that emerge mid-career for

Figure 5.6
Comparison of Predicted DoD Civilian and Private-Sector Pay Trajectories for Cyber Operator in the Washington, D.C. Local Pay Area

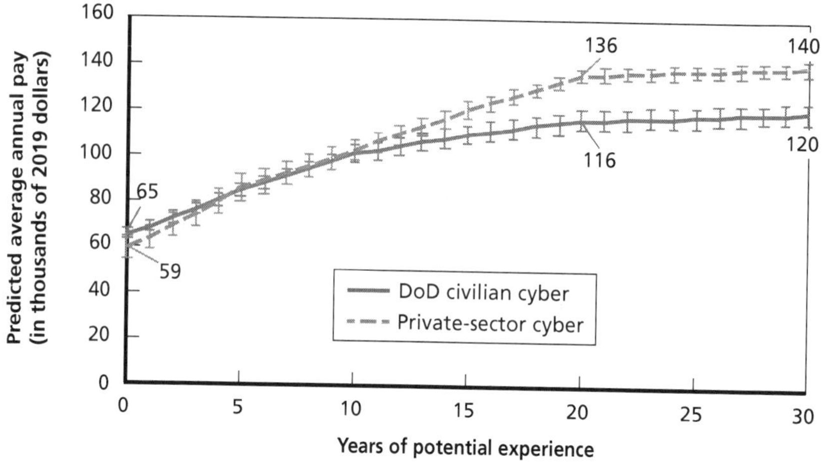

NOTES: Predictions are generated by regression models that account for gender, LPA, educational attainment, and years of potential experience using 2012–2018 ACS and September 2018 DMDC data. Predictions are for men with bachelor's degrees in the Washington, D.C. LPA. Error bars correspond to the prediction's 95 percent confidence interval.

Software Developer (621) and Cyber Operator (321) are large (roughly $12,000 and $20,000, respectively), and they persist through 30 years of potential experience.

Finally, we demonstrate that the pay trajectory differences vary considerably among the 13 LPAs we explored. Cross-LPA results are summarized in Table 5.3 for Cyber Defense Analyst (511), which also reflects Security Control Assessor (612), Systems Security Analyst (461), and Cyber Defense Incident Responder (531). Detailed DoD civilian and private-sector pay trajectories for the 13 LPAs are available in Appendix C.

For three LPAs—Hawaii, Huntsville, and San Diego—the pay trajectories follow similar patterns to that of the Washington, D.C. LPA. There is a DoD civilian pay premium at low years of potential experience (cells colored green) that wanes as the years of potential experience increase (cells colored light green to yellow). In contrast, one

Table 5.3
Cross-LPA Differences in Predicted Private-Sector Pay and Predicted DoD Civilian Pay for Cyber Defense Analyst (in thousands of 2019 dollars)

	Years of Potential Experience						
LPA	**1**	**5**	**10**	**15**	**20**	**25**	**30**
Washington, D.C.	–14	–11	–12	–5	3	2	2
Hawaii	–14	–12	–14	–8	–2	–3	–3
Huntsville	–15	–13	–15	–10	–4	–4	–4
Indianapolis	–5	–1	0	7	14	14	14
Los Angeles	–6	–1	–1	7	14	14	14
New York	–1	5	7	16	25	25	25
Philadelphia	–11	–8	–8	–2	6	6	5
Sacramento	–3	2	3	11	19	19	19
San Diego	–12	–9	–9	–3	5	5	4
San Francisco	3	11	15	26	37	37	37
Seattle	19	30	37	49	61	61	62
St. Louis	–9	–6	–6	0	7	7	6
Tucson	–15	–14	–16	–11	–6	–6	–6

NOTES: Values in the table are in thousands of 2019 dollars and represent the difference in predicted pay of private-sector employees less predicted pay of DoD civilian employees for workers in a Cyber Defense Analyst (511) work role. As cells become greener, the differential is more in favor of DoD civilian employees. As cells become redder, the differential is more in favor of private-sector employees. Predictions are for men with bachelor's degrees.

LPA—Seattle—has dramatically different results from all other LPAs, where the private-sector pay trajectory exceeds the DoD civilian pay trajectory for our seven cyber work roles. For the remaining eight LPAs, the work roles largely follow the same patterns seen in the Washington, D.C. LPA, but they sort differently across work roles in each locality. For instance, in Indianapolis, New York, and Sacramento, the pay trajectories for Cyber Defense Analyst (511), Security Control Assessor (612), Systems Security Analyst (461), and Cyber Defense Incident

Responder (531) follow the pattern that Cyber Operator (321) took in the Washington, D.C. LPA (as shown in Figure 5.6), in which pay is similar in early years and diverges substantially later on. In Los Angeles, Philadelphia, and St. Louis, the pay trajectories for Cyber Defense Analyst (511), Security Control Assessor (612), Systems Security Analyst (461), and Cyber Defense Incident Responder (531) follow the pattern that Software Developer (621) took in the Washington, D.C. LPA (as shown in Figure 5.5), in which there is a DoD civilian pay premium at hiring, but that shrinks as years of potential experience increase until a private-sector pay premium emerges.

In summary, there is noteworthy variation in the DoD civilian and private-sector cyber work role wage trajectories by locality that should not be overlooked. Differences in predicted pay by LPA may reflect the unique features of the local cyber market, whereas federal locality pay is set based on comparable pay for the broader federal workforce. In setting CES pay, adjustments outside of the Washington, D.C. LPA may merit structural adjustments if additional data on labor demand can be produced. Notably, cyber pay differences are substantial between the DoD and the private sector in Seattle (see Table 5.3 and Appendix C) and, to a lesser degree, in several other LPAs (e.g., San Francisco, New York, Indianapolis, Los Angeles, and Sacramento).

CHAPTER SIX

Conclusions, Considerations, and Recommendations

Conclusions

The CES was created to attract and retain high-caliber personnel critical to the DoD cyber warfare mission. In creating the CES, Congress granted the DoD CIO, in conjunction with the USD(P&R), certain compensation flexibilities. Key differences from existing compensation flexibilities include a separate pay schedule, the option to add two steps onto a pay grade, and the ability of the DoD CIO, in conjunction with USD(P&R), to establish TLMSs for specific subpopulations where labor market competitiveness issues are impeding recruiting and retaining the cyber workforce. The main innovation of the CES is placing the responsibility for compensation policy with the DoD CIO and USD(P&R) rather than with OPM. To justify the use of its new compensation flexibilities, notably the TLMS, the DoD CIO needs to document cyber work role shortages in CES-covered agencies. Justification for use of broad pay flexibilities should include

> detailed analysis of recruiting or retention issues regarding the targeted occupational or specialty groups, and supporting evidence that other actions within the existing CES policy framework, including recruitment, relocation, and retention incentives are insufficient to ensure successful maintenance of the required workforce (DoDI 1400.25, Vol. 3006, 2017, p. 10).

Documentation of persistent vacancies and high turnover in a position demonstrates a shortage of labor and represents the strongest case for adjusting pay for that position. Pay adjustments can be temporary or permanent. Changes to base compensation, such as a TLMS, are considered permanent changes to compensation. Targeted incentives, such as recruiting and retention incentives, are not permanent changes to compensation. The use of these incentives is less costly. Only when incentives are insufficient to achieve recruiting and retention goals should permanent income changes be tried.

In this report, we have analyzed the labor demand and supply of seven DoD cyber work roles identified as high priority collectively by the service components and the Office of the DoD CIO. We worked with this office to collect information about employment, vacancies, turnover, and compensation from DoD organizations that have converted or will convert to the CES in the near future. Responses were intended to identify cyber labor shortages and document the use of existing compensation flexibilities to address those shortages.

As discussed in Chapter Three, responses were often incomplete and inconsistent. Consequently, there is insufficient information to identify broad labor shortages in the CES workforce. To justify a permanent pay adjustment (e.g., TLMS), there should be evidence that temporary compensation changes, such as recruiting and retention incentives, are insufficient to attract and retain qualified employees. We saw little evidence that recruiting and retention incentives are being widely used, either in the responses provided by CES organizations or in our independent analysis of data on the DoD civilian cyber workforce's pay. We did identify specific cases in which evidence suggests high turnover and substantial vacancies indicative of a labor shortage and cases in which substantial differences in pay exist. In these cases, the use of compensation flexibilities may be warranted.

A limitation of our analysis is that we only consider market-based reasons for permanent compensation adjustments. It was outside the scope of our analysis to consider issues of compensation equity with other federal cyber employers (e.g., the Intelligence Community) or issues associated with mission risk (e.g., insufficient or unqualified personnel but imminent need). These might provide separate, non–market-based rationales for permanent pay adjustments.

Looking to the future, it is critical that CES organizations systematically collect and track information about employment, vacancies, turnover, and compensation, as highlighted in Chapter Three. This information is necessary to facilitate the DoD CIO in making a strong market-based case for TLMS.

Considerations

The DoD has many available options to try to attract and retain cybersecurity workers. One strategy could be to increase wage offers to particular occupations within a local market. For already-established cybersecurity labor markets (e.g., San Francisco, Seattle), wages might need to increase more than in other metropolitan areas. Wages must be comparable to similarly defined occupations in the private sector for the DoD to compete in these geographical areas. For geographical areas with thin cyber labor markets (i.e., few cybersecurity workers and employers), it is not clear that increasing wages would be the most efficient way to attract and retain a work force. Cybersecurity workers might choose to live in a noncyber hub for reasons of particular preference, such as being close to family or a relatively low cost of living. This labor force might be less responsive to changes in wage offers.

Another potential mechanism to attract and retain civilian cyber workers is to offer attractive nonwage benefits, such as better quality and lower-cost employer-sponsored health insurance, flexible work schedules, and opportunities for continued training. Unfortunately, little data exist on how cybersecurity workers value these types of nonpecuniary compensation mechanisms. There are only a handful of studies that focus on the effects of nonpecuniary benefits on labor supply, and the studies that do exist typically focus on the labor supply of married women. These studies have shown that having a wife with health insurance reduces husbands' labor force participation rate by four to nine percentage points (Wellington and Cobb-Clark, 2000). The link between health insurance and job mobility, however, is still unclear in direction and statistical significance (Gruber and Madrian, 2002). It is established, however, that responsiveness to such nonpecuniary benefits are not uniform across demographics, as the labor supply of women and older workers is more responsive to changes in health insurance coverage and retirement incentives (Blundell, French, and Tetlow, 2016; Buchmueller and Valletta, 1999). Therefore, women and older workers may be more attracted to employment with the DoD if the DoD is able to provide better benefits than comparable private-sector jobs. Another benefit dimension on which the DoD may effectively compete is job stability. Given that the technology sector changes more rapidly than other employment sectors, ensuring stable employment and a positive career trajectory might be attractive to potential employees.

If particular roles are continuously difficult to fill, another potential solution is to modify job market criteria, such as educational requirements at the time of hire or the necessity of being a U.S. citizen. If a particular set of technical skills are rare in the labor market, it might be in the DoD's interest to hire an entry-level individual and invest in training them or to hire foreign students that desire to stay in the United States. Because occupational licenses and certifications create friction in the labor market, decreasing such requirements at the time of hire will widen the net of potential applicants for a particular position.

However, all of the potential solutions discussed so far are blunt instruments; they must be applied across the board to all employees of a certain type in a certain geography. Because labor markets are dynamic and information is not perfect, a more incremental approach might help the DoD achieve its goals at lower cost. For example, the use of hiring or retention bonuses for particularly skilled or high-value workers could be used, together with data on hiring and retention outcomes, work performance, and vacancy levels, to figure out the appropriate competitive salary for a given position in a given area. This has the added benefit of reducing overall costs because targeted incentives do not have to be provided to all employees equally, as do salary or benefit increases. Once the right salary level is discovered, it could be rolled out more generally to all those in a given position and area.

An incremental approach can also be usefully applied to salary and benefit increases. For example, salaries in a given work role in a given geography could be increased a little bit each year over a several year period to observe their effect on intended outcomes. By continuously observing hiring, vacancy, and retention data, the DoD can discover what level of salary is necessary to achieve its goals. For example, if a $2,000 salary increase is slated for two years from now, but it is clear that the last increase was adequate to meet targets, then the planned increase can be discontinued. Similarly, if targets are still not met after several years of small increases, the subsequent increase can be increased. In this way, managers could slowly change bonuses, awards, and salaries over time and each time observe hiring and retention outcomes to better adjust future increases until indicators reach the desired level.

A final strategy is to look for alternatives to permanent workers, such as short-term contractors or firms that provide outsourcing services. Most major technology firms make abundant use of short-term contract labor, allowing them to expand and contract their workforce as needed to accomplish projects. This also allows some companies to better screen for potential future permanent employees in a relatively low-cost way. If a contractor is very productive, that contractor could be converted to a full-time employee at the end of the contract; those who are not good workers would not have their contracts renewed. This trial period provides much more information about a potential employee than the typical hiring process at most firms. When a bad hiring decision is made, the costs can be borne by the firm for decades in terms of lost employee and team productivity. Although hiring contractors is common in cyber firms, it might be difficult for the DoD to do so, given the required security clearance and the necessary time, effort, and cost involved in obtaining those clearances. However, it might be possible to identify tasks or projects that could be outsourced easily or given to a noncleared temporary contract worker.

Recommendations

We detail several recommendations aimed at supporting the DoD CIO in setting compensation policy. These recommendations address the perceived need for pay adjustments in the competitive cyber labor market while aligning incentives of the CES organizations to collect and track the information necessary for the DoD CIO to provide market-based justifications for use of the compensation flexibilities it has been granted.

Continue to Categorize Cyber Personnel by Cyber Work Roles

Many CES organizations seemed unable to identify specific cyber work roles as part of their responses to the Office of the DoD CIO's employment and compensation questionnaire. We recommend that DoD civilian cyber positions be continuously categorized and recorded in the standard record-keeping systems (e.g., civilian personnel records) available to the CES organizations and their human resources offices. This will provide a reference point for human resources managers when asked questions as part of an annual employment and compensation questionnaire. Furthermore, once positions and personnel are consistently categorized, hiring and retention can be monitored using administrative data systems, automating an important part of the business case for TLMSs.

Regularly Collect Data on Employment and Compensation of DoD Civilian Cyber Positions

For justifying and setting TLMSs, quantitative evidence must be regularly collected to demonstrate cyber workforce labor shortages, use of temporary pays to alleviate labor shortages, and the magnitude of pay adjustments required.

To identify and document the existence of labor market shortages by work role, the DoD CIO should annually collect information from CES organizations on employment, vacancies, job postings, offers, acceptances, and hires in specific work roles. Ideally, collection of this information would become routinized through the civilian human resources organizations and available directly to the CES organization and the DoD CIO.[1]

---

[1] An alternative data collection approach may be possible through collaboration with the Defense Civilian Personnel Advisory Service or DMDC. As long as information on employment, vacancies, job postings, offers, acceptances, and hires in specific work roles for the CES organizations can be regularly and accurately collected, the organization doing the collection does not matter. However, collaboration with the Defense Civilian Personnel Advisory Service Wage Division on the design and methodology for conducting salary surveys could increase the accuracy and reliability of the data obtained.

To document use of temporary pays to recruit and retain workers, the DoD CIO should annually collect information from CES organizations on the use of special pays, particularly the use of recruiting and retention incentives. In doing so, the data should focus on offers of the incentives and track acceptance. Existing pay data can only track workers who accepted these offers and started or continued employment. Ideally, this information on compensation can be collected as part of the same annual employer survey used to collect information on employment.

Finally, once labor shortages are identified, the DoD CIO will need to identify the magnitude of pay adjustments required. To support pay setting, the DoD CIO should regularly update information on workforce comparisons between the DoD and the private sector, like those documented in Chapter Five.

Establish a TLMS Adjustment Schedule Based on Verifiable Benchmarks

The DoD CIO should define a plan that can be shared with policymakers, CES organizations, and the CES workforce regarding how it intends to adjust cyber worker pay over the next five years to respond to perceived labor shortages. A clearly articulated plan can assist retention by setting expectations of future pay changes that are different from the rest of the federal workforce. Such a plan should

- use verifiable and work-role-specific benchmarks calculated using administrative data (e.g., ratio of net hires to positions, DoD civilian-to-private-sector pay comparisons)
- include a formulaic adjustment to pay based on labor shortfalls (e.g., TLMS will be adjusted one percentage point per year for every 10 percent of positions that go unfilled)
- have senior leader buy-in that is communicated to the workforce.

This plan makes plain how important it is for CES organizations to collect and track employment and pay data so that the DoD CIO can demonstrate the need for the TLMS. Furthermore, communicating the plan to the workforce will demonstrate the value of the CES to DoD civilian cyber workers.

Consider Structural Wage Adjustments Using TLMS for Work Roles with Major Salary Differences and Existing Labor Shortages

We identified a limited number of specific cases where evidence suggests high turnover and substantial vacancies indicative of a labor shortage and a few cases where there exist substantial differences in pay. In these cases, the use of compensation flexibilities may be warranted.

One specific example was the Cyber Operator (321) work role. Pay for Cyber Operators in the Washington, D.C. LPA is 86 percent of comparable private-sector pay. Evidence from MARFORCYBER, located in this LPA, indicates significant attrition in this work role. Since Cyber Operator is a priority work role for the DoD CIO, an immediate adjustment in pay in the Washington, D.C. LPA (as compared with the formulaic adjustment over time) may be warranted despite any evidence on MARFORCYBER's ability or inability to replace these positions.

The remaining six priority cyber work roles (i.e., 511, 612, 621, 461, 611, 531) receive pay that is 92 percent or more of comparable private-sector work roles for individuals with 20 or more years of potential experience in the Washington, D.C. LPA. In these cases, pay increases should be done incrementally over several years to avoid overadjustment.

Adjustments outside of the Washington, D.C. LPA may merit structural adjustments if additional data on labor demand can be produced. Notably, cyber pay differences are substantial between the DoD and the private sector in Seattle (see Appendix C) and, to a lesser degree, in several other LPAs (e.g., San Francisco, New York, Indianapolis, and Sacramento).

APPENDIX A

Data Sources

In this appendix, we describe the various data sources used for this project. These include the NICE and O*NET data used to create crosswalks between CES cyber work roles and private-sector occupations (see Chapter Four) and the DMDC and ACS PUMS data used to compare pay and other characteristics for DoD and private-sector cyber workers (see Chapter Five). Finally, we use OES data to estimate the potential extent of underreporting of wages in the ACS data.

The NICE Cybersecurity Workforce Framework

NIST published the NICE Cybersecurity Workforce Framework to establish a taxonomy and common lexicon to describe cybersecurity work regardless of where or for whom the work is performed (National Initiative for Cybersecurity Careers and Studies, undated). The NICE Framework facilitates the use of a consistent, comparable, and repeatable approach to selecting cybersecurity roles for particular positions within an organization; provides a common lexicon that academic and governmental institutions can use to develop cybersecurity curricula; and helps employers in the selection of job candidates and development of relevant training opportunities for their workforce.

The NICE Framework creates a roadmap to describe cybersecurity work in the public sector, private sector, and academic sector. Any cybersecurity job or position can be described by identifying the relevant material from one or more components of the NICE framework (Newhouse et al., 2017). Cybersecurity jobs are broken down into categories and specialty areas based on the type of work performed. For example, *cyber investigation* and *digital forensics* are two specialty areas that fall under the *investigate* category, based on similarities in job descriptions. The NICE Framework also nests specific cyber work roles within the categories and specialty areas. These work roles—for example, Cyber Defense Analyst—are assigned a unique work role identifier. Each work role lists the associated KSATs necessary to perform this work role, which are also given a code categorization (National Initiative for Cybersecurity Careers and Studies, undated).

The NICE Framework is therefore a resource for describing and understanding not only the cybersecurity work roles but also the KSATs needed to perform these work roles. Broadly speaking, the NICE Framework helps users communicate about how to identify, recruit, develop, and retain cybersecurity talent.

O*NET Database

O*NET is a primary source of occupational information. The program also owns and operates the O*NET Database, which contains information on hundreds of standardized and occupation-specific descriptors. Researchers continually update the database by conducting surveys with analysts and industrial and organizational psychologists, the results of which are made available to users through the web-based application, O*NET OnLine. O*NET OnLine can be searched by users and provides the basis for career exploration tools that help workers and students looking to find or change careers.

Similar to the work roles identified in the NICE Framework, each occupation within the O*NET database requires a variety of activities and tasks, which are performed using a mix of KSAs. Distinguishing characteristics for each occupation are described by the O*NET Content Model, which identifies key features of an occupation using a standardized and measurable set of variables called *descriptors*. The content model begins with six domains that describe the day-to-day aspects of a job and the qualifications of a typical worker (for example, worker characteristics, worker requirements, experience requirements, occupational requirements, workforce characteristics, occupation-specific information). The model then expands to include the 277 descriptors collected by the O*NET program and more collected by other federal agencies, including the BLS. The content model allows users to identify the different levels of specificity needed for their query.

In addition to identifying characteristics for a single occupation, O*NET OnLine allows users to define sets of occupations using the SOC taxonomy. This taxonomy includes 974 occupations with associated data rated by analysis and tasks that are developed by industrial and organizational psychologists. The data collection program connects the content model, which outlines the information collected, with the O*NET-SOC taxonomy, which defines occupations (O*NET Online, undated a).

DMDC Data

We use data provided by the DMDC to capture information on the characteristics and wages of DoD civilian workers in the CMF and the CPF. The CMF contains demographic information about civilian DoD workers, including birth date, gender, education level, occupation, and grade. We use quarterly records from 2010 through the first quarter of 2017, after which the record updates occurred monthly (from April 2017 through September 2019). CPF updates are also provided quarterly and include information for each pay period. We have data from March 2010 to September 2019. Among other things, these files record base compensation, local market pay supplements, LPA, and incentives and other rewards. Finally, the Office of the DoD CIO provided a list of cyber workers and their work roles mapped to the NICE framework as of June 2019. For more details on how these data are used for the analyses in Chapter Five, see Appendix B.

ACS PUMS

Data on private-sector workers and their wages come from the ACS, an annual survey administered by the U.S. Census Bureau. The survey is conducted on an ongoing basis and updated yearly. The ACS PUMS contains tabulated records on both housing units and individual people. These data provide information on numerous variables of interest, including geographic location, demographics, educational attainment, and earnings.

The ACS identifies occupations using SOC, a coding system used by federal agencies to classify workers into occupational categories. Within this coding system, all workers are categorized into one of the 23 major groups, 98 minor groups, 459 broad occupations, and 867 detailed occupations. Through 2017, the ACS used the 2010 SOC system. A new SOC system was released in 2018. We create a crosswalk to consistently code our occupations of interest in the ACS from 2012 to 2018.

OES

The BLS OES program produces estimates of employment and wages for specific occupations based on a semiannual survey. In addition, the program collects data on wage and salary workers to generate employment and wage estimates for approximately 800 occupations. The program produces these estimates for the nation, by state, and by metropolitan and nonmetropolitan area, as well as by industry and ownership (e.g., sole proprietors).

The OES, which surveys approximately 180,000 to 200,000 establishments per panel, is a federal-state cooperative program between the BLS and state workforce agencies. Employment data are benchmarked to an average of the May and November employment levels; the BLS also uses data collected from the SOC system. The survey covers all full-time and part-time wage and salary workers in nonfarm industries and does not include self-employed workers, owners in unincorporated firms, household workers, or unpaid family workers.

Potential Underreporting of Wages in the ACS

Several researchers have documented that individuals may underreport wages on surveys (see, for example, Moore, Stinson, and Welniak, 2000). Because wages in the ACS are self-reported, underreporting is a potential risk. This might bias private-sector earnings trajectories lower, reducing differences between DoD civilians and private-sector civilians. In contrast, because the OES is derived from administrative data provided by employers, we would not expect it to suffer from underreporting. OES data alone were not sufficient for this study, since we needed information on individual characteristics that are not included in OES data (e.g., age, educational attainment). However, we compared data on wages from the OES and the ACS to explore the potential extent of underreporting in the ACS.

An important feature to note is the different subject samples between the ACS and the OES. The ACS surveys individuals, while the OES surveys establishments. Furthermore, the OES survey does not include the self-employed, owners and partners in unincorporated firms, household workers, or unpaid family workers. Although the ACS data can be restricted to more closely match the OES data (by dropping self-employed and unpaid family workers, for example), estimates between the two measures may differ even in the absence of underreporting, given the different samples and survey methodologies.

In Figure A.1, we plot the two-digit SOC occupation level difference in median wages between the OES and ACS. The figure shows that median wages in the OES are higher than median wages in the ACS, consistent with underreporting in the ACS.

Differences in the ACS and OES are not constant across all wage levels and are greater at higher wages. Figure A.2 depicts a scatter plot of median wages reported in the one-year ACS samples collected during 2012 through 2017 and in the OES for the same years at the two-digit SOC occupation level. If the ACS and OES reported exactly the same median wages for all 23 major occupation codes, then the scatterplot would perfectly follow the 45-degree red line. We see that more points appear above the red line as the median wage increases, which indicates

Figure A.1
Median Wage Differential Among Occupations, OES Versus ACS

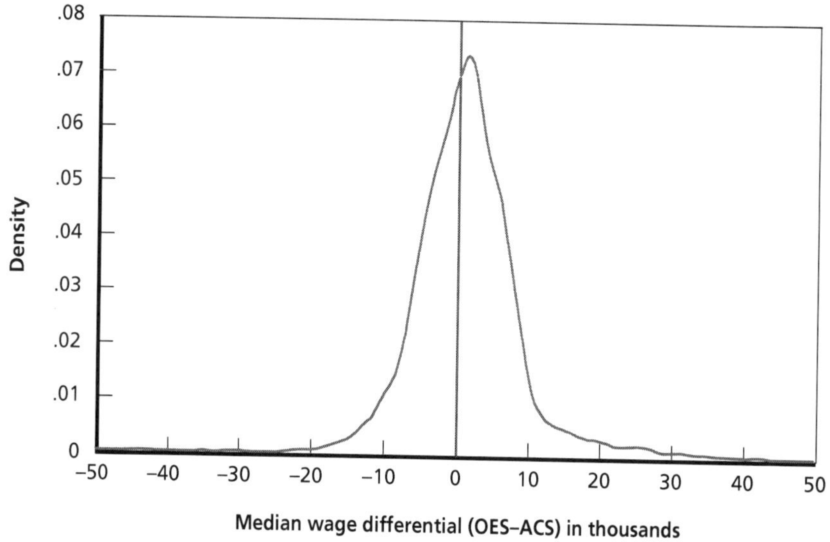

NOTE: Occupational statistics are computed at the two-digit SOC code level and include part-time and full-time workers.

that the difference in median wages between the OES (on the *y*-axis) and the ACS (on the *x*-axis) grows with median wages.

To investigate how the potential underreporting in the ACS might affect the results of our analyses, we compare median and mean wages in the OES and ACS for the specific roles of interest in this study over time. Table A.1 reports the ratio of the OES median and mean wage to the ACS median and mean wage for the work roles of interest for 2018.

As can be seen in Table A.1, median wages in the OES are between 2 percent and 33 percent higher than in the ACS for our work roles of interest, and there is wide variation in the extent of the difference by work role. An argument could be made for adjusting the private-sector pay trajectories reported in Chapter Five upward to account for the potential underreporting of wages in the ACS. For example, wages of Cyber Operators (321) could be adjusted upward by 10 percent, further exacerbating the DoD–private-sector pay gap. We note this potential underreporting for the interested reader. However, we

Figure A.2
Comparison of OES Median Wages to ACS Median Wages

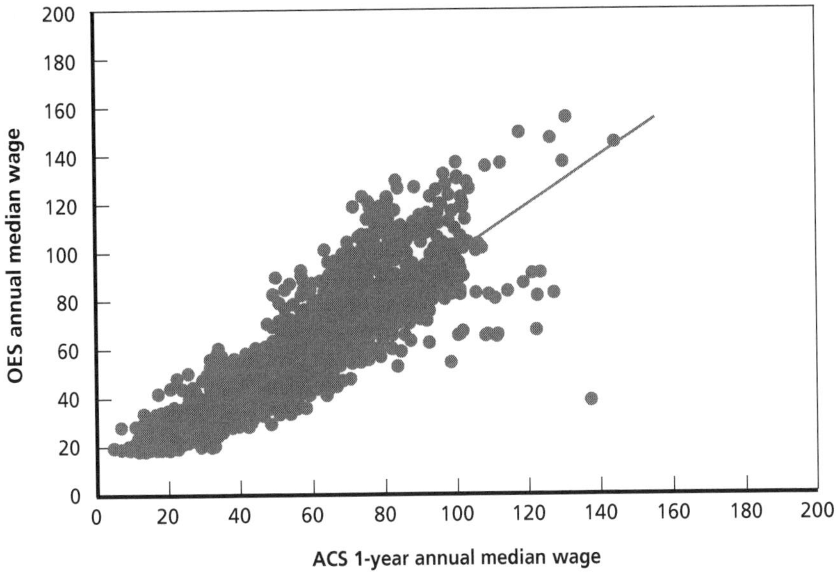

NOTE: Occupational statistics are computed at the two-digit SOC code level and include part-time and full-time workers.

Table A.1
Ratio of OES to ACS Median Wages for Work Roles of Interest, 2018

Cyber Work Role	Ratio of OES-to-ACS Median Wage	Ratio of OES-to-ACS Mean Wage
511, 612, 461, 531	1.0229	0.9606
321	1.1047	1.0711
621	1.0628	0.9611
611	1.3340	1.2755

decided against adjusting the ACS pay values to address underreporting because pay discrepancies are typically insufficient to justify a pay adjustment as discussed in Chapter One, and therefore any adjustment for underreporting would not qualitatively change the implications for our recommendations.

APPENDIX B

Technical Details

In this appendix, we describe several technical details in this report, including

1. the text clustering algorithm used in Chapter Four
2. the process of identifying our sample of civilians in the DoD cyber workforce using DMDC data (this sample was used in Chapter Five)
3. the econometric models we used for estimating pay trajectories for DoD civilian cyber workers using the DMDC data and for private-sector workers using the ACS data (these econometric models were used in the analysis presented in Chapter Five)
4. the regression results from Chapter Five.

Text Clustering Algorithm

Our text clustering algorithm proceeded as follows: For the description of each NICE work role and the set of O*NET jobs in the four categories of interest, we split the descriptions into tokens, or lists of respective words, and their stems, using a Python function called Snowball Stemmer (Snowball, undated). We also used the Natural Language Toolkit (NLTK Project, undated) list of English stop words to take out words that do not convey meaning, such as *a* and *the*. We then created a term frequency-inverse document frequency (tf-idf) matrix, which counts word occurrences by description. This was transformed into a document-term matrix, or term frequency matrix, which allowed us to apply a weighting so that words that occur frequently within a description but infrequently across all descriptions receive a higher weighting; these words help to better define the differences between job descriptions. Using this matrix of weighted word counts, we employed a k-means clustering algorithm. K-means clustering first assigns each description to a cluster to minimize the within cluster sum of squares. The algorithm next takes the mean of the clustered observations to use as the new cluster centroid. Descriptions are then iteratively reassigned to clusters and centroids recalculated until the algorithm convergences. We set the algorithm to divide the data (one NICE work role of interest plus O*NET jobs in the four categories) into ten clusters based on text similarity. We then took the cluster that contained the NICE work role of interest and the O*NET jobs that were in its cluster and passed them on to our SMEs to refine into a best match.

Identifying a Sample of Civilians in the DoD Cyber Workforce and Other Data Standardization Procedures

The Office of the DoD CIO provided us with a list of work roles of interest based on its internal analyses. The office also provided us with a list of workers who were classified as cyber workers as of June 2019, along with their work roles as of that date. Using a unique person-level identifier, we merged this file to the September 2018 and June 2019 DMDC CMF to identify cyber workers. We excluded individuals who switched into one of our work roles of interest between September 2018 and June 2019 (defined as any individual in one of our work roles of interest in June 2019 who had a different occupation code listed in the September 2018 and June 2019 files). We were unable to identify cyber workers in the September 2018 file who left the DoD or switched out of the cyber workforce between September 2018 and June 2019. Our resulting sample of cyber workers includes those who work in one of the cyber work roles of interest and other cyber workers. Workers in other occupations are excluded from the sample.

We then merged these data to the DMDC CPF for the third quarter of 2018 and kept the last pay period in September. To focus on full-time workers, we dropped individuals who worked fewer than 70 hours for the two-week pay period. We define total annual pay as *basic salary*, which records the yearly salary of the individual, plus the product of *locality pay* multiplied by 100.[1]

Information on gender, date of birth, education, and occupation came from the CMF, and information on annual pay and location came from the CPF. We created age from the date of birth variable. We recategorized education into three categories: those with less than a bachelor's degree, those with a bachelor's degree, and those with a master's degree or more. Work role categorizations for personnel come from the data provided by the Office of the DoD CIO. Using the location variable from the CPF, we identified the applicable LPA. We also inflated annual salaries to 2019 dollars using the Consumer Price Index for All Urban Consumers (CPI-U; BLS, 2020b).

DMDC data do not include years of work experience, but we are interested in modeling pay trajectories over an individual's career. To overcome this limitation, we used a common substitute, years of potential experience, which is calculated as the difference between an individual's age and an assumed age at the time they completed their education (as measured before our recategorization). In other words, years of potential experience is calculated as age minus 18 years for high school graduates or high school dropouts, age minus 20 years for those with some college or associate's degrees, age minus 22 years for college graduates, and age minus 24 years for those with advanced degrees (Smith, Asch, and Mattock, 2020). The resulting number is the presumed number of years the individual has been in the workforce.

Pay Trajectory Specification for DoD Civilian Cyber Workers

To compare pay across DoD civilians and private-sector workers in the cyber work roles of interest, we used the following regression equation (Equation B.1) using ordinary least squares:

---

1 Locality pay in the CPFs is listed in each pay period as the annual amount divided by 100.

$$\ln(pay_i) = \alpha + \Sigma_{k=1}^{10}\beta_k Workrole_{ik} + \gamma Male_i + \Sigma_{l=1}^{3}\delta_l Edu_{il} + \Sigma_{m=1}^{47}\Theta_m LPA_{im} + \varphi_1 YOE + \varphi_2[YOE-5]*I_{YOE>5} + \varphi_3[YOE-10]*I_{YOE>10} + \varphi_4[YOE-15]*I_{YOE>15} + \varphi_5[YOE-20]*I_{YOE>20} + \varepsilon_i \quad \text{Eq. B.1}$$

where $\ln(pay_i)$ is the natural log of pay for individual i, each of the *Workrole* variables is a dummy recording whether the individual is in that work role,[2] *Male* records if the individual is male, each *Edu* variable is a dummy recording whether the individual has that level of education, *LPA* is a dummy variable recording whether the individual lives in that particular LPA, *YOE* is years of potential experience as defined above, $I_{YOE>X}$ is a dummy variable that is zero if *YOE* is less than *X* and one otherwise, and ε_i is an error term. Treating the *YOE* variable in this way creates a piecewise function, which helps to better fit the data and better smooth the resulting pay estimates. Standard errors were clustered by work role.

Coefficient estimates from this regression were stored and then used to predict the pay for a male in each work role for the full set of years of potential experience, education categories, and LPAs. These pay estimates were graphed to create the pay trajectories displayed in Chapter Five for DoD civilian cyber workers by work role and education level.

Pay Trajectory Specification for Private-Sector Cyber Workers

We downloaded data for the entire United States from the one-year ACS PUMS for 2012 through 2018 from the U.S. Census Bureau web-

---

[2] Note that besides the seven work roles of interest, we also estimated pay trajectories for Exploitation Analyst (121), System Testing and Evaluation Specialist (671), and Other Cyber Workers, giving us ten work role categories.

site (U.S. Census Bureau, 2020). For a measure of annual pay, we used the variable defined as wages or salary income over the past 12 months. We adjusted this variable to account for the fact that individuals take the survey at different points within the survey time period.[3] We coded the work roles of interest in the ACS based on the occupation SOC as described in Chapter Four. As in the DoD data, we used the CPI-U to adjust all years' wages to 2019 dollars.

We restricted the data to only full-time, employed workers. We dropped those who were unemployed, not in the labor force, self-employed, or working without pay in a family business or farm. We also dropped individuals who worked less than 35 hours per week or less than 27 weeks per year and individuals identified as local, state, or federal government employees.

Occupation SOC codes changed in the 2018 ACS sample, so we mapped our 2012 through 2017 SOC codes of interest to their 2018 equivalents. Cyber workers were identified as those with SOC codes that begin with "1511" from 2012 through 2017 and "1512" in 2018.[4] As with the DMDC data, we dropped workers who were not cyber workers.

To compare ACS and DMDC data, we modified geographic identifiers in the ACS. Public Use Microdata Areas (PUMAs) are the most detailed geographic areas available in the publicly available ACS data. PUMAs are nonoverlapping areas that partition each U.S. state into areas containing approximately 100,000 residents. PUMA definitions are updated every ten years with the decennial census. The 2010 PUMA boundaries were first used in the 2012 ACS data; they cover the entire period of our analyses. We used a PUMA-to-LPA crosswalk to convert PUMAs to LPAs, which are identified in the DoD data. Since PUMAs and LPAs do not overlap completely, we duplicated ACS observations whose PUMAs span LPA boundaries and

---

3 We multiplied pay by "adjinc"/1,000,000 as recommended in ACS documentation. Note that this adjustment makes no difference in our analysis.

4 Note that one of our work roles of interest has an SOC code that does not fall within the cyber worker definition used here. Those in the work role of Authorizing Official (611) are matched with Computer and Information Systems Manager (11-3021).

weight the observations according to the percentage of the PUMA's population that falls within the given LPA. We multiplied this weight with the sample weight provided in the ACS to create new frequency weights (rounded), which we use to weight our pay regressions to be better representative of the civilian population in a given area.

Mirroring the analysis for DoD civilian cyber workers, we estimated the following pay regression using weighted ordinary least squares in Equation B.2:

$$\ln(pay_i) = \alpha + \Sigma_{k=1}^{10} \beta_k Workrole_{ik} + \gamma Male_i + \Sigma_{l=1}^{3} \delta_l Edu_{il} + \Sigma_{m=1}^{47} \Theta_m LPA_{im} + \varphi_1 YOE + \varphi_2 [YOE - 5] * I_{YOE>5} + \varphi_3 [YOE - 10] * I_{YOE>10} + \varphi_4 [YOE - 15] * I_{YOE>15} + \varphi_5 [YOE - 20] * + I_{YOE>20} + \Sigma_{2012}^{2018} \omega_n Year_{in} + \varepsilon_i \qquad \text{Eq. B.2}$$

where, as before, $\ln(Pay_i)$ is the natural log of the pay for individual i, each of the *Workrole* variables is a dummy recording whether the individual is in that work role,[5] *Male* records whether the individual is male, each *Edu* variable is a dummy recording whether the individual has that level of education, *LPA* is a dummy variable recording whether the individual lives in that particular LPA, *YOE* is years of potential experience, $I_{YOE>X}$ is a dummy variable that is zero if *YOE* is less than X and one otherwise, and ε_i is an error term. Unlike in the DoD regression equation, we include data from multiple years (2012 through 2018) in these regressions to increase the sample size and the precisions of the estimates. Therefore, we also include *Year* dummies in the ACS regression. Standard errors are clustered by work role.

As with the DMDC analysis, we generate pay trajectories to compare to those of DoD civilian cyber workers by work role using the

5 Note that multiple DoD work roles are mapped to the same ACS work role, so there are seven work roles in this regression rather than the ten in the DoD regressions. Specifically, 511, 612, 461, and 531 are all mapped to SOC 15-1122 (15-1212 in 2018). As before, we also estimate pay trajectories for Exploitation Analyst (121), System Testing and Evaluation Specialist (671), and Other Cyber Workers, giving us seven work role categories.

coefficient estimates from this regression to estimate the pay for a male in 2018 in each work role for the full set of years of experience, education categories, and LPAs. These pay estimates are graphed to create the pay trajectories displayed in Chapter Five for private-sector cyber workers by work role and education level.

Regression Results

Table B.1 details the results of the pay regressions for government cyber workers and for private-sector cyber workers. Standard errors are in parentheses.

Table B.1
Log Wage Regression Results

Variable	DoD	Private Sector
Work role [Baseline: other cyber workers]		
Cyber Defense Analyst (511)	0.0644*** (0.00115)	0.203*** (0.00241)
Cyber Operator (321)	−0.0258*** (0.00195)	0.250*** (0.00200)
Exploitation Analyst (121)	0.139*** (0.000750)	0.113*** (0.00658)
Security Control Assessor (612)	0.0786*** (0.00147)	
Software Developer (621)	0.0464*** (0.00638)	0.255*** (0.00441)
System Testing and Evaluation Specialist (671)	0.0673*** (0.00469)	0.160*** (0.00704)
Systems Security Analyst (461)	0.0730*** (0.00102)	
Authorizing Official/Designating Representative (611)	0.169*** (0.00338)	0.288*** (0.0105)
Cyber Defense Incident Responder (531)	0.0791*** (0.00137)	

Table B.1—Continued

Variable	DoD	Private Sector
Gender [Baseline: Male]		
Female	0.0208* (0.00769)	−0.155*** (0.00707)
LPA [Baseline: Alaska]		
Albany	0.259 (0.155)	0.0116 (0.0267)
Albuquerque	0.876*** (0.0273)	−0.0536 (0.0703)
Atlanta	0.935*** (0.0245)	0.0710 (0.0381)
Austin	0.813*** (0.0378)	0.122** (0.0269)
Boston	0.987*** (0.0295)	0.177** (0.0351)
Buffalo	0.810*** (0.0272)	−0.0565* (0.0222)
Chicago	0.925*** (0.0318)	0.131** (0.0319)
Cincinnati	0.941*** (0.0269)	0.0425 (0.0393)
Cleveland	0.817*** (0.0489)	−0.0120 (0.0409)
Columbus	0.919*** (0.0488)	0.0462 (0.0390)
Colorado Springs	0.830*** (0.0258)	0.0300 (0.0328)
Charlotte	0.772*** (0.0241)	0.106* (0.0329)
Dayton	0.728*** (0.0138)	−0.0440 (0.0374)
Washington, D.C.	1.102*** (0.0317)	0.218*** (0.0274)
Denver	0.953*** (0.0442)	0.125* (0.0349)

Table B.1—Continued

Variable	DoD	Private Sector
Detroit	0.924*** (0.0298)	0.0109 (0.0354)
Dallas	0.843*** (0.0217)	0.0989* (0.0324)
Davenport	0.866*** (0.0242)	−0.0104 (0.0624)
Hartford	0.921*** (0.0236)	0.111* (0.0398)
Harrisburg	0.868*** (0.0336)	0.0293 (0.0364)
Hawaii	0.916*** (0.0256)	−0.0107 (0.0467)
Huntsville	0.960*** (0.0341)	0.0234 (0.0374)
Houston	0.919*** (0.0420)	0.118* (0.0423)
Indianapolis	0.768*** (0.0176)	0.00331 (0.0364)
Kansas City	0.778*** (0.0316)	0.00286 (0.0356)
Los Angeles	0.901*** (0.0198)	0.126** (0.0283)
Laredo	—	−0.508 (0.216)
Las Vegas	0.769*** (0.0218)	−0.0119 (0.0387)
Miami	0.928*** (0.0400)	−0.0249 (0.0274)
Milwaukee	0.865*** (0.0267)	0.0301 (0.0410)
Minneapolis	0.861*** (0.0204)	0.106* (0.0395)
New York	0.927*** (0.0295)	0.235*** (0.0371)

Table B.1—Continued

Variable	DoD	Private Sector
Palm Bay	0.723*** (0.0141)	0.0560 (0.0454)
Philadelphia	0.963*** (0.0171)	0.110* (0.0369)
Pittsburgh	0.819*** (0.0258)	0.0165 (0.0254)
Portland	0.894*** (0.0451)	0.0885 (0.0450)
Phoenix	0.746*** (0.0201)	0.0515 (0.0285)
Raleigh	0.721*** (0.0228)	0.0940* (0.0255)
Richmond	0.925*** (0.0248)	0.113 (0.0471)
Rest of United States	0.760*** (0.0167)	−0.0410 (0.0310)
Sacramento	0.830*** (0.0610)	0.100* (0.0279)
San Diego	1.026*** (0.0196)	0.162*** (0.0249)
Seattle	0.674*** (0.0230)	0.319*** (0.0529)
San Jose	1.026*** (0.0257)	0.395*** (0.0304)
St. Louis	0.887*** (0.0304)	0.0478 (0.0240)
Tucson	0.881*** (0.0207)	−0.0791 (0.0457)
Education [Baseline: Less than a college degree]		
College grad	0.246*** (0.0187)	0.281*** (0.0250)
Master's plus	0.345*** (0.0232)	0.418*** (0.0416)

Table B.1—Continued

Variable	DoD	Private Sector
YOE [years of experience] Spline		
YOE1	0.0525*** (0.000899)	0.0722*** (0.00225)
YOE2	−0.0180*** (0.00143)	−0.0355*** (0.00200)
YOE3	−0.0168*** (0.000964)	−0.00446* (0.00143)
YOE4	−0.00618*** (0.000583)	−0.00766** (0.00185)
YOE5	−0.00824*** (0.000600)	−0.0219*** (0.00256)
Year [Baseline: 2012]		
2013	—	−0.00720* (0.00235)
2014	—	−0.00713 (0.00482)
2015	—	0.00813 (0.00497)
2016	—	0.0295 (0.0150)
2017	—	0.0381 −(0.0166)
2018	—	0.0386 (0.0237)
Constant	9.761*** (0.0347)	10.21*** (0.0475)
Observations	50,963	25,513,343
Adj. R-Sq	0.096	0.336

NOTES: In the ACS data, the following work roles were coded to the same SOC code, and the result of the regression for these roles appears in the first row: Cyber Defense Analyst (511), Security Control Assessor (612), Systems Security Analyst (461), Cyber Defense Incident Responder (531). The constant captures the baseline (excluded categories) of other cyber workers, males, with less than a college degree, in Alaska, in 2012. Standard errors are shown in parentheses. * $p < 0.05$, ** $p < 0.01$, *** $p < 0.001$.

APPENDIX C

Detailed Pay Trajectory Data by Locality

In the main body of the report, we presented pay trajectory results graphically for the Washington, D.C. LPA. In this appendix, we provide the detailed pay trajectory estimates for the Washington, D.C. LPA and the following other 12 localities:

1. Hawaii
2. Huntsville
3. Indianapolis
4. Los Angeles
5. New York
6. Philadelphia
7. Sacramento
8. San Diego
9. San Francisco
10. Seattle
11. St. Louis
12. Tucson.

These 13 localities were selected because a nontrivial share (at least 5 percent) of workers in a cyber work role are in these localities.

In predicting these pay trajectories by location, work role, and years of potential experience, we hold other worker characteristics constant. These characteristics include education, gender, and year. Tables C.1 through C.13 provide predictions for 2018 male workers with a bachelor's degree. The salary amounts are presented in thousands of dollars.

Table C.1
Predicted Average Annual Pay for DoD Civilian and Private-Sector Cyber Work Roles in Washington, D.C. LPA

	Authorizing Official (611)		Cyber Defense Analyst (511)		Cyber Defense Incident Responder (531)		Cyber Operator (321)		Security Control Assessor (612)		Software Developer (621)		Systems Security Analyst (461)	
YOPE	DoD	Private	DoD	Private	DoD	Private	DoD	Private	DoD	Private	DoD	Private	DoD	Private
0	$79.1	$61.8	$71.2	$56.8	$72.3	$56.8	$65.1	$59.5	$72.2	$56.8	$69.9	$59.8	$71.8	$56.8
1	$83.4	$66.4	$75.0	$61.0	$76.2	$61.0	$68.6	$63.9	$76.1	$61.0	$73.7	$64.3	$75.7	$61.0
2	$87.8	$71.4	$79.1	$65.6	$80.3	$65.6	$72.3	$68.7	$80.2	$65.6	$77.7	$69.1	$79.8	$65.6
3	$92.6	$76.7	$83.4	$70.5	$84.6	$70.5	$75.2	$73.9	$84.5	$70.5	$81.9	$74.2	$84.1	$70.5
4	$97.6	$82.4	$87.8	$75.8	$89.1	$75.8	$80.3	$79.4	$89.1	$75.8	$86.3	$79.8	$88.6	$75.8
5	$102.8	$88.6	$92.6	$81.4	$94.0	$81.4	$84.6	$85.3	$93.9	$81.4	$90.9	$85.8	$93.4	$81.4
6	$106.4	$91.9	$95.8	$84.5	$97.3	$84.5	$87.6	$88.5	$97.2	$84.5	$94.1	$89.0	$96.7	$84.5
7	$110.2	$95.4	$99.2	$87.6	$100.7	$87.6	$90.6	$91.8	$100.6	$87.6	$97.4	$92.3	$100.1	$87.6
8	$114.0	$98.9	$102.7	$90.9	$104.2	$90.9	$93.8	$95.2	$104.1	$90.9	$100.8	$95.7	$103.6	$90.9
9	$118.0	$102.6	$106.3	$94.3	$107.8	$94.3	$97.1	$98.8	$107.8	$94.3	$104.4	$99.3	$107.2	$94.3
10	$122.2	$106.4	$110.0	$97.8	$111.6	$97.8	$100.5	$102.5	$111.6	$97.8	$108.0	$103.0	$111.0	$97.8
11	$124.4	$109.9	$112.0	$101.0	$113.6	$101.0	$102.3	$105.8	$113.6	$101.0	$110.0	$106.4	$112.9	$101.0
12	$126.6	$113.5	$114.0	$104.3	$115.6	$104.3	$104.1	$109.3	$115.6	$104.3	$111.9	$109.9	$114.9	$104.3
13	$128.8	$117.2	$116.0	$107.7	$117.7	$107.7	$106.0	$112.9	$117.6	$107.7	$113.9	$113.5	$117.0	$107.7

Table C.1—continued

	Authorizing Official (611)		Cyber Defense Analyst (511)		Cyber Defense Incident Responder (531)		Cyber Operator (321)		Security Control Assessor (612)		Software Developer (621)		Systems Security Analyst (461)	
YOPE	DoD	Private	DoD	Private	DoD	Private	DoD	Private	DoD	Private	DoD	Private	DoD	Private
14	$131.1	$121.1	$118.1	$111.2	$119.8	$111.2	$107.9	$116.6	$119.7	$111.2	$116.0	$117.2	$119.1	$111.2
15	$133.5	$125.0	$120.2	$114.9	$121.9	$114.9	$109.8	$120.4	$121.9	$114.9	$118.0	$121.0	$121.2	$114.9
16	$135.0	$128.1	$121.5	$117.7	$123.3	$117.7	$111.1	$123.4	$123.3	$117.7	$119.4	$124.0	$122.6	$117.7
17	$136.6	$131.3	$122.9	$120.7	$124.8	$120.7	$112.3	$126.4	$124.7	$120.7	$120.8	$127.1	$124.0	$120.7
18	$138.1	$134.6	$124.4	$123.7	$126.2	$123.7	$113.6	$129.6	$126.1	$123.7	$122.2	$130.2	$125.4	$123.7
19	$139.7	$137.9	$125.8	$126.7	$127.7	$126.7	$114.9	$132.8	$127.6	$126.7	$123.6	$133.5	$126.9	$126.7
20	$141.4	$141.3	$127.3	$129.9	$129.1	$129.9	$116.3	$136.1	$129.1	$129.9	$125.0	$136.8	$128.4	$129.9
21	$141.8	$141.7	$127.7	$130.2	$129.6	$130.2	$116.7	$136.5	$129.5	$130.2	$125.4	$137.2	$128.8	$130.2
22	$142.3	$142.1	$128.1	$130.6	$130.0	$130.6	$117.0	$136.8	$129.9	$130.6	$125.8	$137.5	$129.2	$130.6
23	$142.7	$142.5	$128.5	$130.9	$130.4	$130.9	$117.4	$137.2	$130.3	$130.9	$126.2	$137.9	$129.6	$130.9
24	$143.2	$142.9	$128.9	$131.3	$130.8	$131.3	$117.8	$137.6	$130.8	$131.3	$126.6	$138.3	$130.0	$131.3
25	$143.7	$143.2	$129.3	$131.6	$131.3	$131.6	$118.2	$137.9	$131.2	$131.6	$127.0	$138.6	$130.5	$131.6
26	$144.1	$143.6	$129.8	$132.0	$131.7	$132.0	$118.6	$138.3	$131.6	$132.0	$127.4	$139.0	$130.9	$132.0
27	$144.6	$144.0	$130.2	$132.3	$132.1	$132.3	$118.9	$138.7	$132.0	$132.3	$127.9	$139.4	$131.3	$132.3

Table C.1—continued

	Authorizing Official (611)		Cyber Defense Analyst (511)		Cyber Defense Incident Responder (531)		Cyber Operator (321)		Security Control Assessor (612)		Software Developer (621)		Systems Security Analyst (461)	
YOPE	DoD	Private	DoD	Private	DoD	Private	DoD	Private	DoD	Private	DoD	Private	DoD	Private
28	$145.1	$144.4	$130.6	$132.7	$132.5	$132.7	$119.3	$139.0	$132.5	$132.7	$128.3	$139.7	$131.7	$132.7
29	$145.5	$144.8	$131.0	$133.0	$133.0	$133.0	$119.7	$139.4	$132.9	$133.0	$128.7	$140.1	$132.2	$133.0
30	$146.0	$145.2	$131.4	$133.4	$133.4	$133.4	$120.1	$139.8	$133.3	$133.4	$129.1	$140.5	$132.6	$133.4

NOTES: Values are predictions from a regression model that accounts for years of potential experience, education, gender, and work role using data from the 2012–2018 ACS and the September 2018 DMDC CMF. Predictions reflect 2018 male workers with a bachelor's degree. YOPE = years of potential experience.

Table C.2
Predicted Average Annual Pay for DoD Civilian and Private-Sector Cyber Work Roles in Hawaii LPA

	Authorizing Official (611)		Cyber Defense Analyst (511)		Cyber Defense Incident Responder (531)		Cyber Operator (321)		Security Control Assessor (612)		Software Developer (621)		Systems Security Analyst (461)	
YOPE	DoD	Private	DoD	Private	DoD	Private	DoD	Private	DoD	Private	DoD	Private	DoD	Private
0	$65.7	$49.2	$59.1	$45.2	$60.0	$45.2	$54.0	$47.3	$60.0	$45.2	$58.1	$47.6	$59.7	$45.2
1	$69.2	$52.8	$62.3	$48.5	$63.3	$48.5	$57.0	$50.9	$63.2	$48.5	$61.2	$51.1	$62.9	$48.5
2	$73.0	$56.8	$65.7	$52.2	$66.7	$52.2	$60.0	$54.7	$66.6	$52.2	$64.5	$55.0	$66.3	$52.2
3	$76.9	$61.0	$69.2	$56.1	$70.3	$56.1	$63.3	$58.8	$70.2	$56.1	$68.0	$59.1	$69.8	$56.1
4	$81.0	$65.6	$73.0	$60.3	$74.0	$60.3	$66.7	$63.2	$74.0	$60.3	$71.7	$63.5	$73.6	$60.3
5	$85.4	$70.5	$76.9	$64.8	$78.0	$64.8	$70.3	$67.9	$78.0	$64.8	$75.5	$68.2	$77.6	$64.8
6	$88.4	$73.1	$79.6	$67.2	$80.8	$67.2	$72.7	$70.4	$80.7	$67.2	$78.2	$70.8	$80.3	$67.2
7	$91.5	$75.9	$82.4	$69.7	$83.6	$69.7	$75.3	$73.1	$83.6	$69.7	$80.9	$73.4	$83.1	$69.7
8	$94.7	$78.7	$85.3	$72.3	$86.5	$72.3	$77.9	$75.8	$86.5	$72.3	$83.8	$76.2	$86.0	$72.3
9	$98.0	$81.6	$88.3	$75.0	$89.6	$75.0	$80.7	$78.6	$89.5	$75.0	$86.7	$79.0	$89.0	$75.0
10	$101.5	$84.7	$91.4	$77.8	$92.7	$77.8	$83.5	$81.5	$92.7	$77.8	$89.7	$82.0	$92.2	$77.8
11	$103.3	$87.5	$93.0	$80.4	$94.4	$80.4	$85.0	$84.2	$94.3	$80.4	$91.3	$84.6	$93.8	$80.4
12	$105.1	$90.3	$94.7	$83.0	$96.1	$83.0	$86.5	$87.0	$96.0	$83.0	$93.0	$87.4	$95.5	$83.0
13	$107.0	$93.3	$96.3	$85.7	$97.8	$85.7	$88.0	$89.8	$97.7	$85.7	$94.6	$90.3	$97.2	$85.7

Table C.2—continued

	Authorizing Official (611)		Cyber Defense Analyst (511)		Cyber Defense Incident Responder (531)		Cyber Operator (321)		Security Control Assessor (612)		Software Developer (621)		Systems Security Analyst (461)	
YOPE	DoD	Private	DoD	Private	DoD	Private	DoD	Private	DoD	Private	DoD	Private	DoD	Private
14	$108.9	$96.3	$98.1	$88.5	$99.5	$88.5	$89.6	$92.8	$99.5	$88.5	$96.3	$93.2	$98.9	$88.5
15	$110.9	$99.5	$99.8	$91.4	$101.3	$91.4	$91.2	$95.8	$101.2	$91.4	$98.0	$96.3	$100.7	$91.4
16	$112.1	$101.9	$101.0	$93.7	$102.4	$93.7	$92.2	$98.2	$102.4	$93.7	$99.2	$98.7	$101.8	$93.7
17	$113.4	$104.5	$102.1	$96.0	$103.6	$96.0	$93.3	$100.6	$103.6	$96.0	$100.3	$101.1	$103.0	$96.0
18	$114.7	$107.1	$103.3	$98.4	$104.8	$98.4	$94.4	$103.1	$104.8	$98.4	$101.5	$103.6	$104.2	$98.4
19	$116.1	$109.7	$104.5	$100.8	$106.0	$100.8	$95.5	$105.7	$106.0	$100.8	$102.6	$106.2	$105.4	$100.8
20	$117.4	$112.5	$105.7	$103.3	$107.3	$103.3	$96.6	$108.3	$107.2	$103.3	$103.8	$108.8	$106.6	$103.3
21	$117.8	$112.8	$106.0	$103.6	$107.6	$103.6	$96.9	$108.6	$107.6	$103.6	$104.2	$109.1	$107.0	$103.6
22	$118.2	$113.1	$106.4	$103.9	$108.0	$103.9	$97.2	$108.9	$107.9	$103.9	$104.5	$109.4	$107.3	$103.9
23	$118.6	$113.4	$106.7	$104.2	$108.3	$104.2	$97.5	$109.2	$108.3	$104.2	$104.8	$109.7	$107.7	$104.2
24	$118.9	$113.7	$107.1	$104.4	$108.7	$104.4	$97.8	$109.4	$108.6	$104.4	$105.2	$110.0	$108.0	$104.4
25	$119.3	$114.0	$107.4	$104.7	$109.0	$104.7	$98.2	$109.7	$109.0	$104.7	$105.5	$110.3	$108.4	$104.7
26	$119.7	$114.3	$107.8	$105.0	$109.4	$105.0	$98.5	$110.0	$109.3	$105.0	$105.9	$110.6	$108.7	$105.0
27	$120.1	$114.6	$108.1	$105.3	$109.7	$105.3	$98.8	$110.3	$109.7	$105.3	$106.2	$110.9	$109.1	$105.3

Table C.2—continued

YOPE	Authorizing Official (611)		Cyber Defense Analyst (511)		Cyber Defense Incident Responder (531)		Cyber Operator (321)		Security Control Assessor (612)		Software Developer (621)		Systems Security Analyst (461)	
	DoD	Private	DoD	Private	DoD	Private	DoD	Private	DoD	Private	DoD	Private	DoD	Private
28	$120.5	$114.9	$108.5	$105.6	$110.1	$105.6	$99.1	$110.6	$110.0	$105.6	$106.5	$111.2	$109.4	$105.6
29	$120.9	$115.2	$108.8	$105.8	$110.4	$105.8	$99.4	$110.9	$110.4	$105.8	$106.9	$111.5	$109.8	$105.8
30	$121.3	$115.5	$109.2	$106.1	$110.8	$106.1	$99.8	$111.2	$110.7	$106.1	$107.2	$111.8	$110.1	$106.1

NOTES: Values are predictions from a regression model that accounts for years of potential experience, education, gender, and work role using data from the 2012–2018 ACS and the September 2018 DMDC CMF. Predictions reflect 2018 male workers with a bachelor's degree. YOPE = years of potential experience.

Table C.3
Predicted Average Annual Pay for DoD Civilian and Private-Sector Cyber Work Roles in Huntsville LPA

	Authorizing Official (611)		Cyber Defense Analyst (511)		Cyber Defense Incident Responder (531)		Cyber Operator (321)		Security Control Assessor (612)		Software Developer (621)		Systems Security Analyst (461)	
YOPE	DoD	Private	DoD	Private	DoD	Private	DoD	Private	DoD	Private	DoD	Private	DoD	Private
0	$68.7	$50.9	$61.8	$46.7	$62.7	$46.7	$56.5	$49.0	$62.7	$46.7	$60.7	$49.2	$62.3	$46.7
1	$72.4	$54.7	$65.1	$50.2	$66.1	$50.2	$59.5	$52.6	$66.1	$50.2	$64.0	$52.9	$65.7	$50.2
2	$76.3	$58.8	$68.7	$54.0	$69.7	$54.0	$62.7	$56.6	$69.6	$54.0	$67.4	$56.9	$69.2	$54.0
3	$80.4	$63.1	$72.4	$58.0	$73.4	$58.0	$66.1	$60.8	$73.4	$58.0	$71.1	$61.1	$73.0	$58.0
4	$84.7	$67.9	$76.3	$62.4	$77.4	$62.4	$69.7	$65.4	$77.3	$62.4	$74.9	$65.7	$76.9	$62.4
5	$89.3	$73.0	$80.4	$67.0	$81.6	$67.0	$73.4	$70.2	$81.5	$67.0	$78.9	$70.6	$81.1	$67.0
6	$92.4	$75.7	$83.2	$69.5	$84.4	$69.5	$76.0	$72.9	$84.4	$69.5	$81.7	$73.2	$83.9	$69.5
7	$95.6	$78.5	$86.1	$72.1	$87.4	$72.1	$78.7	$75.6	$87.3	$72.1	$84.6	$76.0	$86.9	$72.1
8	$99.0	$81.4	$89.1	$74.8	$90.4	$74.8	$81.4	$78.4	$90.4	$74.8	$87.5	$78.8	$89.9	$74.8
9	$102.5	$84.5	$92.3	$77.6	$93.6	$77.6	$84.3	$81.3	$93.6	$77.6	$90.6	$81.7	$93.1	$77.6
10	$106.1	$87.6	$95.5	$80.5	$96.9	$80.5	$87.2	$84.4	$96.9	$80.5	$93.8	$84.8	$96.3	$80.5
11	$108.0	$90.5	$97.2	$83.2	$98.6	$83.2	$88.8	$87.1	$98.6	$83.2	$95.5	$87.6	$98.0	$83.2
12	$109.9	$93.4	$98.9	$85.9	$100.4	$85.9	$90.4	$90.0	$100.3	$85.9	$97.2	$90.4	$99.8	$85.9
13	$111.8	$96.5	$100.7	$88.7	$102.2	$88.7	$92.0	$92.9	$102.1	$88.7	$98.9	$93.4	$101.6	$88.7

Table C.3—continued

	Authorizing Official (611)		Cyber Defense Analyst (511)		Cyber Defense Incident Responder (531)		Cyber Operator (321)		Security Control Assessor (612)		Software Developer (621)		Systems Security Analyst (461)	
YOPE	DoD	Private	DoD	Private	DoD	Private	DoD	Private	DoD	Private	DoD	Private	DoD	Private
14	$113.8	$99.7	$102.5	$91.6	$104.0	$91.6	$93.6	$96.0	$103.9	$91.6	$100.7	$96.5	$103.4	$91.6
15	$115.9	$102.9	$104.3	$94.6	$105.8	$94.6	$95.3	$99.1	$105.8	$94.6	$102.4	$99.6	$105.2	$94.6
16	$117.2	$105.5	$105.5	$96.9	$107.1	$96.9	$96.4	$101.6	$107.0	$96.9	$103.6	$102.1	$106.4	$96.9
17	$118.5	$108.1	$106.7	$99.3	$108.3	$99.3	$97.5	$104.1	$108.3	$99.3	$104.8	$104.6	$107.7	$99.3
18	$119.9	$110.8	$108.0	$101.8	$109.6	$101.8	$98.6	$106.7	$109.5	$101.8	$106.0	$107.2	$108.9	$101.8
19	$121.3	$113.5	$109.2	$104.3	$110.8	$104.3	$99.8	$109.3	$110.8	$104.3	$107.3	$109.9	$110.2	$104.3
20	$122.7	$116.4	$110.5	$106.9	$112.1	$106.9	$100.9	$112.0	$112.0	$106.9	$108.5	$112.6	$111.4	$106.9
21	$123.1	$116.7	$110.8	$107.2	$112.5	$107.2	$101.3	$112.3	$112.4	$107.2	$108.8	$112.9	$111.8	$107.2
22	$123.5	$117.0	$111.2	$107.5	$112.8	$107.5	$101.6	$112.6	$112.8	$107.5	$109.2	$113.2	$112.1	$107.5
23	$123.9	$117.3	$111.5	$107.8	$113.2	$107.8	$101.9	$112.9	$113.1	$107.8	$109.6	$113.5	$112.5	$107.8
24	$124.3	$117.6	$111.9	$108.1	$113.6	$108.1	$102.2	$113.2	$113.5	$108.1	$109.9	$113.8	$112.9	$108.1
25	$124.7	$117.9	$112.3	$108.4	$113.9	$108.4	$102.6	$113.5	$113.9	$108.4	$110.3	$114.1	$113.2	$108.4
26	$125.1	$118.2	$112.6	$108.6	$114.3	$108.6	$102.9	$113.8	$114.2	$108.6	$110.6	$114.4	$113.6	$108.6
27	$125.5	$118.6	$113.0	$108.9	$114.7	$108.9	$103.2	$114.2	$114.6	$108.9	$111.0	$114.7	$114.0	$108.9

Table C.3—continued

YOPE	Authorizing Official (611)		Cyber Defense Analyst (511)		Cyber Defense Incident Responder (531)		Cyber Operator (321)		Security Control Assessor (612)		Software Developer (621)		Systems Security Analyst (461)	
	DoD	Private	DoD	Private	DoD	Private	DoD	Private	DoD	Private	DoD	Private	DoD	Private
28	$125.9	$118.9	$113.4	$109.2	$115.0	$109.2	$103.6	$114.5	$115.0	$109.2	$111.3	$115.0	$114.4	$109.2
29	$126.3	$119.2	$113.7	$109.5	$115.4	$109.5	$103.9	$114.8	$115.4	$109.5	$111.7	$115.3	$114.7	$109.5
30	$126.7	$119.5	$114.1	$109.8	$115.8	$109.8	$104.3	$115.1	$115.7	$109.8	$112.1	$115.7	$115.1	$109.8

NOTES: Values are predictions from a regression model that accounts for years of potential experience, education, gender, and work role using data from the 2012–2018 ACS and the September 2018 DMDC CMF. Predictions reflect 2018 male workers with a bachelor's degree. YOPE = years of potential experience.

Table C.4.
Predicted Average Annual Pay for DoD Civilian and Private-Sector Cyber Work Roles in Indianapolis LPA

	Authorizing Official (611)		Cyber Defense Analyst (511)		Cyber Defense Incident Responder (531)		Cyber Operator (321)		Security Control Assessor (612)		Software Developer (621)		Systems Security Analyst (461)	
YOPE	DoD	Private	DoD	Private	DoD	Private	DoD	Private	DoD	Private	DoD	Private	DoD	Private
0	$56.6	$49.8	$51.0	$45.8	$51.8	$45.8	$46.6	$48.0	$51.7	$45.8	$50.1	$48.2	$51.4	$45.8
1	$59.7	$53.6	$53.7	$49.2	$54.5	$49.2	$49.1	$51.6	$54.5	$49.2	$52.8	$51.8	$54.2	$49.2
2	$62.9	$57.6	$56.6	$52.9	$57.5	$52.9	$51.8	$55.4	$57.5	$52.9	$55.6	$55.7	$57.1	$52.9
3	$66.3	$61.9	$59.7	$56.9	$60.6	$56.9	$54.6	$59.6	$60.6	$56.9	$58.6	$59.9	$60.2	$56.9
4	$69.9	$66.5	$62.9	$61.1	$63.9	$61.1	$57.5	$64.1	$63.8	$61.1	$61.8	$64.4	$63.5	$61.1
5	$73.7	$71.5	$66.3	$65.7	$67.3	$65.7	$60.6	$68.8	$67.3	$65.7	$65.1	$69.2	$66.9	$65.7
6	$76.2	$74.2	$68.6	$68.2	$69.7	$68.2	$62.7	$71.4	$69.6	$68.2	$67.4	$71.8	$69.2	$68.2
7	$78.9	$76.9	$71.0	$70.7	$72.1	$70.7	$64.9	$74.1	$72.1	$70.7	$69.8	$74.5	$71.7	$70.7
8	$81.7	$79.8	$73.5	$73.3	$74.6	$73.3	$67.2	$76.8	$74.6	$73.3	$72.2	$77.2	$74.2	$73.3
9	$84.6	$82.8	$76.1	$76.1	$77.2	$76.1	$69.6	$79.7	$77.2	$76.1	$74.8	$80.1	$76.8	$76.1
10	$87.5	$85.9	$78.8	$78.9	$80.0	$78.9	$72.0	$82.7	$79.9	$78.9	$77.4	$83.1	$79.5	$78.9
11	$89.1	$88.7	$80.2	$81.5	$81.4	$81.5	$73.3	$85.4	$81.3	$81.5	$78.8	$85.8	$80.9	$81.5
12	$90.7	$91.6	$81.6	$84.2	$82.8	$84.2	$74.6	$88.2	$82.8	$84.2	$80.2	$88.6	$82.3	$84.2
13	$92.3	$94.6	$83.1	$86.9	$84.3	$86.9	$75.9	$91.1	$84.3	$86.9	$81.6	$91.5	$83.8	$86.9

Table C.4—continued

	Authorizing Official (611)		Cyber Defense Analyst (511)		Cyber Defense Incident Responder (531)		Cyber Operator (321)		Security Control Assessor (612)		Software Developer (621)		Systems Security Analyst (461)	
YOPE	DoD	Private	DoD	Private	DoD	Private	DoD	Private	DoD	Private	DoD	Private	DoD	Private
14	$93.9	$97.7	$84.6	$89.8	$85.8	$89.8	$77.3	$94.1	$85.8	$89.8	$83.1	$94.5	$85.3	$89.8
15	$95.6	$100.9	$86.1	$92.7	$87.3	$92.7	$78.6	$97.1	$87.3	$92.7	$84.5	$97.6	$86.8	$92.7
16	$96.7	$103.4	$87.1	$95.0	$88.3	$95.0	$79.5	$99.5	$88.3	$95.0	$85.5	$100.1	$87.8	$95.0
17	$97.8	$105.9	$88.1	$97.4	$89.4	$97.4	$80.5	$102.0	$89.3	$97.4	$86.5	$102.5	$88.8	$97.4
18	$98.9	$108.6	$89.1	$99.8	$90.4	$99.8	$81.4	$104.6	$90.4	$99.8	$87.5	$105.1	$89.9	$99.8
19	$100.1	$111.3	$90.1	$102.3	$91.4	$102.3	$82.3	$107.1	$91.4	$102.3	$88.5	$107.7	$90.9	$102.3
20	$101.2	$114.0	$91.1	$104.8	$92.5	$104.8	$83.3	$109.8	$92.5	$104.8	$89.5	$110.4	$91.9	$104.8
21	$101.6	$114.3	$91.4	$105.1	$92.8	$105.1	$83.6	$110.1	$92.8	$105.1	$89.8	$110.7	$92.2	$105.1
22	$101.9	$114.7	$91.7	$105.4	$93.1	$105.4	$83.8	$110.4	$93.1	$105.4	$90.1	$111.0	$92.5	$105.4
23	$102.2	$115.0	$92.0	$105.6	$93.4	$105.6	$84.1	$110.7	$93.4	$105.6	$90.4	$111.3	$92.8	$105.6
24	$102.6	$115.3	$92.3	$105.9	$93.7	$105.9	$84.4	$111.0	$93.7	$105.9	$90.7	$111.6	$93.1	$105.9
25	$102.9	$115.6	$92.6	$106.2	$94.0	$106.2	$84.6	$111.3	$94.0	$106.2	$91.0	$111.9	$93.4	$106.2
26	$103.2	$115.9	$92.9	$106.5	$94.3	$106.5	$84.9	$111.6	$94.3	$106.5	$91.3	$112.1	$93.7	$106.5
27	$103.6	$116.2	$93.2	$106.8	$94.6	$106.8	$85.2	$111.9	$94.6	$106.8	$91.6	$112.4	$94.1	$106.8

Table C.4—continued

	Authorizing Official (611)		Cyber Defense Analyst (511)		Cyber Defense Incident Responder (531)		Cyber Operator (321)		Security Control Assessor (612)		Software Developer (621)		Systems Security Analyst (461)	
YOPE	DoD	Private	DoD	Private	DoD	Private	DoD	Private	DoD	Private	DoD	Private	DoD	Private
28	$103.9	$116.5	$93.5	$107.1	$94.9	$107.1	$85.5	$112.2	$94.9	$107.1	$91.9	$112.8	$94.4	$107.1
29	$104.2	$116.8	$93.8	$107.3	$95.2	$107.3	$85.7	$112.5	$95.2	$107.3	$92.2	$113.1	$94.7	$107.3
30	$104.6	$117.1	$94.1	$107.6	$95.5	$107.6	$86.0	$112.8	$95.5	$107.6	$92.5	$113.4	$95.0	$107.6

NOTES: Values are predictions from a regression model that accounts for years of potential experience, education, gender, and work role using data from the 2012–2018 ACS and the September 2018 DMDC CMF. Predictions reflect 2018 male workers with a bachelor's degree.
YOPE = years of potential experience.

Table C.5
Predicted Average Annual Pay for DoD Civilian and Private-Sector Cyber Work Roles in Los Angeles LPA

	Authorizing Official (611)		Cyber Defense Analyst (511)		Cyber Defense Incident Responder (531)		Cyber Operator (321)		Security Control Assessor (612)		Software Developer (621)		Systems Security Analyst (461)	
YOPE	DoD	Private	DoD	Private	DoD	Private	DoD	Private	DoD	Private	DoD	Private	DoD	Private
0	$64.7	$56.4	$58.2	$51.8	$59.1	$51.8	$53.2	$54.3	$59.1	$51.8	$57.2	$54.5	$58.8	$51.8
1	$68.2	$60.6	$61.4	$55.7	$62.3	$55.7	$56.1	$58.3	$62.3	$55.7	$60.3	$58.6	$61.9	$55.7
2	$71.9	$65.1	$64.7	$59.8	$65.7	$59.8	$59.1	$62.7	$65.6	$59.8	$63.5	$63.0	$65.3	$59.8
3	$75.7	$70.0	$68.2	$64.3	$69.2	$64.3	$62.3	$67.4	$69.2	$64.3	$67.0	$67.7	$68.8	$64.3
4	$79.8	$75.2	$71.9	$69.1	$72.9	$69.1	$65.7	$72.4	$72.9	$69.1	$70.6	$72.8	$72.5	$69.1
5	$84.1	$80.8	$75.7	$74.3	$76.9	$74.3	$69.2	$77.8	$76.8	$74.3	$74.4	$78.2	$76.4	$74.3
6	$87.1	$83.9	$78.4	$77.0	$79.6	$77.0	$71.6	$80.7	$79.5	$77.0	$77.0	$81.1	$79.1	$77.0
7	$90.1	$87.0	$81.1	$79.9	$82.3	$79.9	$74.1	$83.8	$82.3	$79.9	$79.7	$84.2	$81.9	$79.9
8	$93.3	$90.2	$84.0	$82.9	$85.2	$82.9	$76.7	$86.9	$85.2	$82.9	$82.5	$87.3	$84.7	$82.9
9	$96.6	$93.6	$86.9	$86.0	$88.2	$86.0	$79.4	$90.1	$88.2	$86.0	$85.4	$90.6	$87.7	$86.0
10	$100.0	$97.1	$90.0	$89.2	$91.3	$89.2	$82.2	$93.5	$91.3	$89.2	$88.4	$94.0	$90.8	$89.2
11	$101.7	$100.3	$91.6	$92.1	$92.9	$92.1	$83.7	$96.5	$92.9	$92.1	$90.0	$97.0	$92.4	$92.1
12	$103.5	$103.5	$93.2	$95.1	$94.6	$95.1	$85.2	$99.7	$94.6	$95.1	$91.6	$100.2	$94.0	$95.1
13	$105.4	$106.9	$94.9	$98.3	$96.3	$98.3	$86.7	$103.0	$96.2	$98.3	$93.2	$103.5	$95.7	$98.3

Table C.5—continued

	Authorizing Official (611)		Cyber Defense Analyst (511)		Cyber Defense Incident Responder (531)		Cyber Operator (321)		Security Control Assessor (612)		Software Developer (621)		Systems Security Analyst (461)	
YOPE	DoD	Private	DoD	Private	DoD	Private	DoD	Private	DoD	Private	DoD	Private	DoD	Private
14	$107.3	$110.4	$96.6	$101.5	$98.0	$101.5	$88.2	$106.3	$98.0	$101.5	$94.9	$106.9	$97.4	$101.5
15	$109.2	$114.0	$98.3	$104.8	$99.7	$104.8	$89.8	$109.8	$99.7	$104.8	$96.5	$110.4	$99.1	$104.8
16	$110.4	$116.9	$99.4	$107.4	$100.9	$107.4	$90.8	$112.5	$100.9	$107.4	$97.7	$113.1	$100.3	$107.4
17	$111.7	$119.8	$100.6	$110.1	$102.1	$110.1	$91.9	$115.3	$102.0	$110.1	$98.8	$115.9	$101.5	$110.1
18	$113.0	$122.8	$101.7	$112.8	$103.2	$112.8	$93.0	$118.2	$103.2	$112.8	$99.9	$118.8	$102.6	$112.8
19	$114.3	$125.8	$102.9	$115.6	$104.4	$115.6	$94.0	$121.1	$104.4	$115.6	$101.1	$121.8	$103.8	$115.6
20	$115.6	$128.9	$104.1	$118.5	$105.6	$118.5	$95.1	$124.1	$105.6	$118.5	$102.2	$124.8	$105.0	$118.5
21	$116.0	$129.3	$104.4	$118.8	$106.0	$118.8	$95.4	$124.5	$105.9	$118.8	$102.6	$125.1	$105.3	$118.8
22	$116.4	$129.6	$104.8	$119.1	$106.3	$119.1	$95.7	$124.8	$106.3	$119.1	$102.9	$125.4	$105.7	$119.1
23	$116.8	$130.0	$105.1	$119.4	$106.7	$119.4	$96.0	$125.1	$106.6	$119.4	$103.2	$125.8	$106.0	$119.4
24	$117.1	$130.3	$105.5	$119.7	$107.0	$119.7	$96.4	$125.5	$107.0	$119.7	$103.6	$126.1	$106.4	$119.7
25	$117.5	$130.7	$105.8	$120.1	$107.4	$120.1	$96.7	$125.8	$107.3	$120.1	$103.9	$126.5	$106.7	$120.1
26	$117.9	$131.0	$106.1	$120.4	$107.7	$120.4	$97.0	$126.1	$107.7	$120.4	$104.3	$126.8	$107.1	$120.4
27	$118.3	$131.4	$106.5	$120.7	$108.1	$120.7	$97.3	$126.5	$108.0	$120.7	$104.6	$127.1	$107.4	$120.7

Table C.5—continued

	Authorizing Official (611)		Cyber Defense Analyst (511)		Cyber Defense Incident Responder (531)		Cyber Operator (321)		Security Control Assessor (612)		Software Developer (621)		Systems Security Analyst (461)	
YOPE	DoD	Private	DoD	Private	DoD	Private	DoD	Private	DoD	Private	DoD	Private	DoD	Private
28	$118.7	$131.7	$106.8	$121.0	$108.4	$121.0	$97.6	$126.8	$108.4	$121.0	$104.9	$127.5	$107.8	$121.0
29	$119.1	$132.1	$107.2	$121.4	$108.8	$121.4	$97.9	$127.2	$108.7	$121.4	$105.3	$127.8	$108.1	$121.4
30	$119.4	$132.4	$107.5	$121.7	$109.1	$121.7	$98.3	$127.5	$109.1	$121.7	$105.6	$128.1	$108.5	$121.7

NOTES: Values are predictions from a regression model that accounts for years of potential experience, education, gender, and work role using data from the 2012–2018 ACS and the September 2018 DMDC CMF. Predictions reflect 2018 male workers with a bachelor's degree. YOPE = years of potential experience.

Table C.6
Predicted Average Annual Pay for DoD Civilian and Private-Sector Cyber Work Roles in New York LPA

	Authorizing Official (611)		Cyber Defense Analyst (511)		Cyber Defense Incident Responder (531)		Cyber Operator (321)		Security Control Assessor (612)		Software Developer (621)		Systems Security Analyst (461)	
YOPE	DoD	Private	DoD	Private	DoD	Private	DoD	Private	DoD	Private	DoD	Private	DoD	Private
0	$66.4	$62.8	$59.8	$57.7	$60.7	$57.7	$54.6	$60.5	$60.6	$57.7	$58.7	$60.8	$60.3	$57.7
1	$70.0	$67.5	$63.0	$62.0	$63.9	$62.0	$57.6	$65.0	$63.9	$62.0	$61.9	$65.3	$63.5	$62.0
2	$73.7	$72.6	$66.4	$66.7	$67.4	$66.7	$60.7	$69.9	$67.3	$66.7	$65.2	$70.2	$67.0	$66.7
3	$77.7	$78.0	$70.0	$71.7	$71.0	$71.7	$63.9	$75.1	$71.0	$71.7	$68.7	$75.5	$70.6	$71.7
4	$81.9	$83.8	$73.7	$77.0	$74.8	$77.0	$67.4	$80.7	$74.8	$77.0	$72.4	$81.1	$74.4	$77.0
5	$86.3	$90.1	$77.7	$82.8	$78.9	$82.8	$71.0	$86.8	$78.8	$82.8	$76.3	$87.2	$78.4	$82.8
6	$89.4	$93.5	$80.4	$85.9	$81.6	$85.9	$73.5	$90.0	$81.6	$85.9	$79.0	$90.5	$81.1	$85.9
7	$92.5	$97.0	$83.3	$89.1	$84.5	$89.1	$76.1	$93.4	$84.5	$89.1	$81.8	$93.8	$84.0	$89.1
8	$95.7	$100.6	$86.2	$92.4	$87.5	$92.4	$78.7	$96.8	$87.4	$92.4	$84.6	$97.3	$86.9	$92.4
9	$99.1	$104.3	$89.2	$95.9	$90.5	$95.9	$81.5	$100.5	$90.5	$95.9	$87.6	$101.0	$90.0	$95.9
10	$102.6	$108.2	$92.3	$99.4	$93.7	$99.4	$84.4	$104.2	$93.7	$99.4	$90.7	$104.7	$93.1	$99.4
11	$104.4	$111.8	$94.0	$102.7	$95.4	$102.7	$85.9	$107.6	$95.3	$102.7	$92.3	$108.2	$94.8	$102.7
12	$106.2	$115.4	$95.7	$106.1	$97.1	$106.1	$87.4	$111.1	$97.0	$106.1	$94.0	$111.7	$96.5	$106.1
13	$108.1	$119.2	$97.4	$109.5	$98.8	$109.5	$89.0	$114.8	$98.8	$109.5	$95.6	$115.4	$98.2	$109.5

Table C.6—continued

	Authorizing Official (611)		Cyber Defense Analyst (511)		Cyber Defense Incident Responder (531)		Cyber Operator (321)		Security Control Assessor (612)		Software Developer (621)		Systems Security Analyst (461)	
YOPE	DoD	Private	DoD	Private	DoD	Private	DoD	Private	DoD	Private	DoD	Private	DoD	Private
14	$110.1	$123.1	$99.1	$113.1	$100.6	$113.1	$90.5	$118.5	$100.5	$113.1	$97.3	$119.1	$100.0	$113.1
15	$112.0	$127.1	$100.9	$116.8	$102.4	$116.8	$92.2	$122.4	$102.3	$116.8	$99.1	$123.0	$101.7	$116.8
16	$113.3	$130.3	$102.0	$119.7	$103.5	$119.7	$93.2	$125.4	$103.5	$119.7	$100.2	$126.1	$102.9	$119.7
17	$114.6	$133.5	$103.2	$122.7	$104.7	$122.7	$94.3	$128.6	$104.7	$122.7	$101.4	$129.2	$104.1	$122.7
18	$116.0	$136.8	$104.4	$125.7	$105.9	$125.7	$95.4	$131.8	$105.9	$125.7	$102.5	$132.4	$105.3	$125.7
19	$117.3	$140.2	$105.6	$128.9	$107.2	$128.9	$96.5	$135.0	$107.1	$128.9	$103.7	$135.7	$106.5	$128.9
20	$118.6	$143.7	$106.8	$132.1	$108.4	$132.1	$97.6	$138.4	$108.3	$132.1	$104.9	$139.1	$107.7	$132.1
21	$119.0	$144.1	$107.2	$132.4	$108.8	$132.4	$97.9	$138.8	$108.7	$132.4	$105.3	$139.5	$108.1	$132.4
22	$119.4	$144.5	$107.5	$132.8	$109.1	$132.8	$98.2	$139.1	$109.1	$132.8	$105.6	$139.8	$108.4	$132.8
23	$119.8	$144.9	$107.9	$133.1	$109.5	$133.1	$98.6	$139.5	$109.4	$133.1	$105.9	$140.2	$108.8	$133.1
24	$120.2	$145.3	$108.2	$133.5	$109.8	$133.5	$98.9	$139.9	$109.8	$133.5	$106.3	$140.6	$109.2	$133.5
25	$120.6	$145.6	$108.6	$133.8	$110.2	$133.8	$99.2	$140.2	$110.1	$133.8	$106.6	$141.0	$109.5	$133.8
26	$121.0	$146.0	$108.9	$134.2	$110.5	$134.2	$99.5	$140.6	$110.5	$134.2	$107.0	$141.3	$109.9	$134.2
27	$121.4	$146.4	$109.3	$134.5	$110.9	$134.5	$99.8	$141.0	$110.8	$134.5	$107.3	$141.7	$110.2	$134.5

Table C.6—continued

	Authorizing Official (611)		Cyber Defense Analyst (511)		Cyber Defense Incident Responder (531)		Cyber Operator (321)		Security Control Assessor (612)		Software Developer (621)		Systems Security Analyst (461)	
YOPE	DoD	Private	DoD	Private	DoD	Private	DoD	Private	DoD	Private	DoD	Private	DoD	Private
28	$121.8	$146.8	$109.6	$134.9	$111.2	$134.9	$100.2	$141.4	$111.2	$134.9	$107.7	$142.1	$110.6	$134.9
29	$122.2	$147.2	$110.0	$135.3	$111.6	$135.3	$100.5	$141.7	$111.6	$135.3	$108.0	$142.5	$110.9	$135.3
30	$122.6	$147.6	$110.3	$135.6	$112.0	$135.6	$100.8	$142.1	$111.9	$135.6	$108.4	$142.8	$111.3	$135.6

NOTES: Values are predictions from a regression model that accounts for years of potential experience, education, gender, and work role using data from the 2012–2018 ACS and the September 2018 DMDC CMF. Predictions reflect 2018 male workers with a bachelor's degree. YOPE = years of potential experience.

Table C.7
Predicted Average Annual Pay for DoD Civilian and Private-Sector Cyber Work Roles in Philadelphia LPA

	Authorizing Official (611)		Cyber Defense Analyst (511)		Cyber Defense Incident Responder (531)		Cyber Operator (321)		Security Control Assessor (612)		Software Developer (621)		Systems Security Analyst (461)	
YOPE	DoD	Private	DoD	Private	DoD	Private	DoD	Private	DoD	Private	DoD	Private	DoD	Private
0	$68.9	$55.5	$62.0	$51.0	$62.9	$51.0	$56.7	$53.4	$62.9	$51.0	$60.9	$53.7	$62.5	$51.0
1	$72.6	$59.6	$65.3	$54.8	$66.3	$54.8	$59.7	$57.4	$66.3	$54.8	$64.2	$57.7	$65.9	$54.8
2	$76.5	$64.1	$68.9	$58.9	$69.9	$58.9	$62.9	$61.7	$69.9	$58.9	$67.6	$62.0	$69.5	$58.9
3	$80.6	$68.9	$72.6	$63.3	$73.7	$63.3	$66.3	$66.3	$73.6	$63.3	$71.3	$66.6	$73.2	$63.3
4	$85.0	$74.0	$76.5	$68.0	$77.6	$68.0	$69.9	$71.3	$77.6	$68.0	$75.1	$71.6	$77.2	$68.0
5	$89.5	$79.6	$80.6	$73.1	$81.8	$73.1	$73.7	$76.6	$81.8	$73.1	$79.2	$77.0	$81.3	$73.1
6	$92.7	$82.5	$83.4	$75.8	$84.7	$75.8	$76.2	$79.5	$84.6	$75.8	$82.0	$79.9	$84.2	$75.8
7	$95.9	$85.6	$86.4	$78.7	$87.7	$78.7	$78.9	$82.4	$87.6	$78.7	$84.8	$82.9	$87.1	$78.7
8	$99.3	$88.8	$89.4	$81.6	$90.7	$81.6	$81.7	$85.5	$90.7	$81.6	$87.8	$85.9	$90.2	$81.6
9	$102.8	$92.1	$92.5	$84.6	$93.9	$84.6	$84.6	$88.7	$93.9	$84.6	$90.9	$89.2	$93.3	$84.6
10	$106.4	$95.6	$95.8	$87.8	$97.2	$87.8	$87.5	$92.0	$97.2	$87.8	$94.1	$92.5	$96.6	$87.8
11	$108.3	$98.7	$97.5	$90.7	$98.9	$90.7	$89.1	$95.0	$98.9	$90.7	$95.8	$95.5	$98.3	$90.7
12	$110.2	$101.9	$99.2	$93.6	$100.7	$93.6	$90.7	$98.1	$100.6	$93.6	$97.5	$98.6	$100.1	$93.6
13	$112.2	$105.2	$101.0	$96.7	$102.5	$96.7	$92.3	$101.3	$102.4	$96.7	$99.2	$101.9	$101.9	$96.7

Table C.7—continued

	Authorizing Official (611)		Cyber Defense Analyst (511)		Cyber Defense Incident Responder (531)		Cyber Operator (321)		Security Control Assessor (612)		Software Developer (621)		Systems Security Analyst (461)	
YOPE	DoD	Private	DoD	Private	DoD	Private	DoD	Private	DoD	Private	DoD	Private	DoD	Private
14	$114.2	$108.7	$102.8	$99.9	$104.3	$99.9	$93.9	$104.7	$104.3	$99.9	$101.0	$105.2	$103.7	$99.9
15	$116.2	$112.2	$104.6	$103.1	$106.2	$103.1	$95.6	$108.1	$106.1	$103.1	$102.8	$108.6	$105.5	$103.1
16	$117.6	$115.0	$105.8	$105.7	$107.4	$105.7	$96.7	$110.8	$107.3	$105.7	$103.9	$111.3	$106.8	$105.7
17	$118.9	$117.9	$107.1	$108.3	$108.6	$108.3	$97.8	$113.5	$108.6	$108.3	$105.1	$114.1	$108.0	$108.3
18	$120.3	$120.8	$108.3	$111.0	$109.9	$111.0	$98.9	$116.3	$109.8	$111.0	$106.4	$116.9	$109.2	$111.0
19	$121.7	$123.8	$109.5	$113.8	$111.2	$113.8	$100.1	$119.2	$111.1	$113.8	$107.6	$119.8	$110.5	$113.8
20	$123.1	$126.9	$110.8	$116.6	$112.4	$116.6	$101.2	$122.2	$112.4	$116.6	$108.8	$122.8	$111.8	$116.6
21	$123.5	$127.2	$111.2	$116.9	$112.8	$116.9	$101.6	$122.5	$112.8	$116.9	$109.2	$123.1	$112.1	$116.9
22	$123.9	$127.6	$111.5	$117.2	$113.2	$117.2	$101.9	$122.8	$113.1	$117.2	$109.5	$123.5	$112.5	$117.2
23	$124.3	$127.9	$111.9	$117.5	$113.5	$117.5	$102.2	$123.2	$113.5	$117.5	$109.9	$123.8	$112.9	$117.5
24	$124.7	$128.3	$112.3	$117.9	$113.9	$117.9	$102.6	$123.5	$113.9	$117.9	$110.3	$124.1	$113.2	$117.9
25	$125.1	$128.6	$112.6	$118.2	$114.3	$118.2	$102.9	$123.8	$114.2	$118.2	$110.6	$124.5	$113.6	$118.2
26	$125.5	$128.9	$113.0	$118.5	$114.7	$118.5	$103.2	$124.2	$114.6	$118.5	$111.0	$124.8	$114.0	$118.5
27	$125.9	$129.3	$113.3	$118.8	$115.0	$118.8	$103.6	$124.5	$115.0	$118.8	$111.3	$125.1	$114.3	$118.8

Table C.7—continued

YOPE	Authorizing Official (611)		Cyber Defense Analyst (511)		Cyber Defense Incident Responder (531)		Cyber Operator (321)		Security Control Assessor (612)		Software Developer (621)		Systems Security Analyst (461)	
	DoD	Private	DoD	Private	DoD	Private	DoD	Private	DoD	Private	DoD	Private	DoD	Private
28	$126.3	$129.6	$113.7	$119.1	$115.4	$119.1	$103.9	$124.8	$115.3	$119.1	$111.7	$125.5	$114.7	$119.1
29	$126.7	$130.0	$114.1	$119.4	$115.8	$119.4	$104.2	$125.2	$115.7	$119.4	$112.1	$125.8	$115.1	$119.4
30	$127.1	$130.3	$114.5	$119.8	$116.2	$119.8	$104.6	$125.5	$116.1	$119.8	$112.4	$126.1	$115.5	$119.8

NOTES: Values are predictions from a regression model that accounts for years of potential experience, education, gender, and work role using data from the 2012–2018 ACS and the September 2018 DMDC CMF. Predictions reflect 2018 male workers with a bachelor's degree. YOPE = years of potential experience.

Table C.8
Predicted Average Annual Pay for DoD Civilian and Private-Sector Cyber Work Roles in Sacramento LPA

	Authorizing Official (611)		Cyber Defense Analyst (511)		Cyber Defense Incident Responder (531)		Cyber Operator (321)		Security Control Assessor (612)		Software Developer (621)		Systems Security Analyst (461)	
YOPE	DoD	Private	DoD	Private	DoD	Private	DoD	Private	DoD	Private	DoD	Private	DoD	Private
0	$60.2	$54.9	$54.2	$50.5	$55.0	$50.5	$49.6	$52.9	$55.0	$50.5	$53.3	$53.2	$54.7	$50.5
1	$63.5	$59.0	$57.2	$54.2	$58.0	$54.2	$52.2	$56.8	$58.0	$54.2	$56.1	$57.1	$57.7	$54.2
2	$66.9	$63.5	$60.2	$58.3	$61.1	$58.3	$55.0	$61.1	$61.1	$58.3	$59.2	$61.4	$60.8	$58.3
3	$70.5	$68.2	$63.5	$62.7	$64.4	$62.7	$58.0	$65.7	$64.4	$62.7	$62.4	$66.0	$64.0	$62.7
4	$74.3	$73.3	$66.9	$67.4	$67.9	$67.4	$61.1	$70.6	$67.9	$67.4	$65.7	$70.9	$67.5	$67.4
5	$78.3	$78.8	$70.5	$72.4	$71.6	$72.4	$64.4	$75.9	$71.5	$72.4	$69.3	$76.3	$71.1	$72.4
6	$81.1	$81.7	$73.0	$75.1	$74.1	$75.1	$66.7	$78.7	$74.0	$75.1	$71.7	$79.1	$73.6	$75.1
7	$83.9	$84.8	$75.6	$77.9	$76.7	$77.9	$69.0	$81.6	$76.6	$77.9	$74.2	$82.0	$76.2	$77.9
8	$86.9	$87.9	$78.2	$80.8	$79.4	$80.8	$71.5	$84.7	$79.3	$80.8	$76.8	$85.1	$78.9	$80.8
9	$89.9	$91.2	$80.9	$83.8	$82.1	$83.8	$74.0	$87.8	$82.1	$83.8	$79.5	$88.3	$81.6	$83.8
10	$93.1	$94.6	$83.8	$87.0	$85.0	$87.0	$76.6	$91.1	$85.0	$87.0	$82.3	$91.6	$84.5	$87.0
11	$94.7	$97.7	$85.3	$89.8	$86.5	$89.8	$77.9	$94.1	$86.5	$89.8	$83.8	$94.6	$86.0	$89.8
12	$96.4	$100.9	$86.8	$92.7	$88.1	$92.7	$79.3	$97.2	$88.0	$92.7	$85.3	$97.7	$87.6	$92.7
13	$98.1	$104.2	$88.3	$95.8	$89.7	$95.8	$80.7	$100.4	$89.6	$95.8	$86.8	$100.9	$89.1	$95.8

Table C.8—continued

	Authorizing Official (611)		Cyber Defense Analyst (511)		Cyber Defense Incident Responder (531)		Cyber Operator (321)		Security Control Assessor (612)		Software Developer (621)		Systems Security Analyst (461)	
YOPE	DoD	Private	DoD	Private	DoD	Private	DoD	Private	DoD	Private	DoD	Private	DoD	Private
14	$99.9	$107.6	$89.9	$98.9	$91.2	$98.9	$82.2	$103.6	$91.2	$98.9	$88.3	$104.2	$90.7	$98.9
15	$101.7	$111.2	$91.5	$102.1	$92.9	$102.1	$83.6	$107.0	$92.8	$102.1	$89.9	$107.6	$92.3	$102.1
16	$102.8	$113.9	$92.6	$104.7	$93.9	$104.7	$84.6	$109.7	$93.9	$104.7	$90.9	$110.2	$93.4	$104.7
17	$104.0	$116.7	$93.6	$107.3	$95.0	$107.3	$85.6	$112.4	$95.0	$107.3	$92.0	$113.0	$94.5	$107.3
18	$105.2	$119.6	$94.7	$109.9	$96.1	$109.9	$86.6	$115.2	$96.1	$109.9	$93.0	$115.8	$95.5	$109.9
19	$106.4	$122.6	$95.8	$112.7	$97.2	$112.7	$87.6	$118.1	$97.2	$112.7	$94.1	$118.7	$96.7	$112.7
20	$107.7	$125.7	$96.9	$115.5	$98.4	$115.5	$88.6	$121.0	$98.3	$115.5	$95.2	$121.6	$97.8	$115.5
21	$108.0	$126.0	$97.2	$115.8	$98.7	$115.8	$88.9	$121.3	$98.6	$115.8	$95.5	$121.9	$98.1	$115.8
22	$108.4	$126.3	$97.6	$116.1	$99.0	$116.1	$89.1	$121.7	$99.0	$116.1	$95.8	$122.3	$98.4	$116.1
23	$108.7	$126.7	$97.9	$116.4	$99.3	$116.4	$89.4	$122.0	$99.3	$116.4	$96.1	$122.6	$98.7	$116.4
24	$109.1	$127.0	$98.2	$116.7	$99.6	$116.7	$89.7	$122.3	$99.6	$116.7	$96.4	$122.9	$99.0	$116.7
25	$109.4	$127.4	$98.5	$117.0	$100.0	$117.0	$90.0	$122.6	$99.9	$117.0	$96.8	$123.3	$99.4	$117.0
26	$109.8	$127.7	$98.8	$117.3	$100.3	$117.3	$90.3	$123.0	$100.2	$117.3	$97.1	$123.6	$99.7	$117.3
27	$110.1	$128.0	$99.1	$117.6	$100.6	$117.6	$90.6	$123.3	$100.6	$117.6	$97.4	$123.9	$100.0	$117.6

Table C.8—continued

YOPE	Authorizing Official (611)		Cyber Defense Analyst (511)		Cyber Defense Incident Responder (531)		Cyber Operator (321)		Security Control Assessor (612)		Software Developer (621)		Systems Security Analyst (461)	
	DoD	Private	DoD	Private	DoD	Private	DoD	Private	DoD	Private	DoD	Private	DoD	Private
28	$110.5	$128.4	$99.5	$118.0	$100.9	$118.0	$90.9	$123.6	$100.9	$118.0	$97.7	$124.2	$100.3	$118.0
29	$110.8	$128.7	$99.8	$118.3	$101.3	$118.3	$91.2	$123.9	$101.2	$118.3	$98.0	$124.6	$100.7	$118.3
30	$111.2	$129.1	$100.1	$118.6	$101.6	$118.6	$91.5	$124.3	$101.6	$118.6	$98.3	$124.9	$101.0	$118.6

NOTES: Values are predictions from a regression model that accounts for years of potential experience, education, gender, and work role using data from the 2012–2018 ACS and the September 2018 DMDC CMF. Predictions reflect 2018 male workers with a bachelor's degree. YOPE = years of potential experience.

Table C.9
Predicted Average Annual Pay for DoD Civilian and Private-Sector Cyber Work Roles in San Diego LPA

	Authorizing Official (611)		Cyber Defense Analyst (511)		Cyber Defense Incident Responder (531)		Cyber Operator (321)		Security Control Assessor (612)		Software Developer (621)		Systems Security Analyst (461)	
YOPE	DoD	Private	DoD	Private	DoD	Private	DoD	Private	DoD	Private	DoD	Private	DoD	Private
0	$73.3	$58.4	$66.0	$53.7	$67.0	$53.7	$60.3	$56.3	$66.9	$53.7	$64.8	$56.5	$66.6	$53.7
1	$77.2	$62.8	$69.5	$57.7	$70.6	$57.7	$63.5	$60.5	$70.5	$57.7	$68.3	$60.8	$70.1	$57.7
2	$81.4	$67.5	$73.3	$62.0	$74.4	$62.0	$67.0	$65.0	$74.3	$62.0	$72.0	$65.3	$73.9	$62.0
3	$85.8	$72.5	$77.2	$66.7	$78.4	$66.7	$70.6	$69.9	$78.3	$66.7	$75.9	$70.2	$77.9	$66.7
4	$90.4	$78.0	$81.4	$71.7	$82.6	$71.7	$74.4	$75.1	$82.6	$71.7	$80.0	$75.5	$82.1	$71.7
5	$95.3	$83.8	$85.8	$77.0	$87.1	$77.0	$78.4	$80.7	$87.0	$77.0	$84.3	$81.1	$86.5	$77.0
6	$98.6	$86.9	$88.8	$79.9	$90.1	$79.9	$81.1	$83.7	$90.1	$79.9	$87.2	$84.1	$89.6	$79.9
7	$102.1	$90.2	$91.9	$82.9	$93.3	$82.9	$84.0	$86.8	$93.2	$82.9	$90.3	$87.3	$92.7	$82.9
8	$105.7	$93.6	$95.1	$86.0	$96.6	$86.0	$86.9	$90.1	$96.5	$86.0	$93.5	$90.5	$96.0	$86.0
9	$109.4	$97.0	$98.5	$89.2	$99.9	$89.2	$90.0	$93.4	$99.9	$89.2	$96.7	$93.9	$99.3	$89.2
10	$113.2	$100.7	$101.9	$92.5	$103.5	$92.5	$93.1	$96.9	$103.4	$92.5	$100.1	$97.4	$102.8	$92.5
11	$115.2	$104.0	$103.8	$95.5	$105.3	$95.5	$94.8	$100.1	$105.2	$95.5	$101.9	$100.6	$104.7	$95.5
12	$117.3	$107.4	$105.6	$98.7	$107.2	$98.7	$96.5	$103.4	$107.1	$98.7	$103.7	$103.9	$106.5	$98.7
13	$119.4	$110.9	$107.5	$101.9	$109.1	$101.9	$98.2	$106.8	$109.0	$101.9	$105.6	$107.3	$108.4	$101.9

Table C.9—continued

	Authorizing Official (611)		Cyber Defense Analyst (511)		Cyber Defense Incident Responder (531)		Cyber Operator (321)		Security Control Assessor (612)		Software Developer (621)		Systems Security Analyst (461)	
YOPE	DoD	Private	DoD	Private	DoD	Private	DoD	Private	DoD	Private	DoD	Private	DoD	Private
14	$121.5	$114.5	$109.4	$105.2	$111.0	$105.2	$100.0	$110.3	$111.0	$105.2	$107.5	$110.8	$110.4	$105.2
15	$123.7	$118.2	$111.3	$108.7	$113.0	$108.7	$101.7	$113.9	$112.9	$108.7	$109.4	$114.4	$112.3	$108.7
16	$125.1	$121.2	$112.6	$111.4	$114.3	$111.4	$102.9	$116.7	$114.2	$111.4	$110.6	$117.3	$113.6	$111.4
17	$126.6	$124.2	$113.9	$114.1	$115.6	$114.1	$104.1	$119.6	$115.6	$114.1	$111.9	$120.2	$114.9	$114.1
18	$128.0	$127.3	$115.3	$117.0	$117.0	$117.0	$105.3	$122.6	$116.9	$117.0	$113.2	$123.2	$116.3	$117.0
19	$129.5	$130.4	$116.6	$119.9	$118.3	$119.9	$106.5	$125.6	$118.3	$119.9	$114.5	$126.2	$117.6	$119.9
20	$131.0	$133.7	$117.9	$122.8	$119.7	$122.8	$107.8	$128.7	$119.6	$122.8	$115.8	$129.4	$119.0	$122.8
21	$131.4	$134.0	$118.3	$123.2	$120.1	$123.2	$108.1	$129.1	$120.0	$123.2	$116.2	$129.7	$119.3	$123.2
22	$131.8	$134.4	$118.7	$123.5	$120.5	$123.5	$108.5	$129.4	$120.4	$123.5	$116.6	$130.1	$119.7	$123.5
23	$132.3	$134.7	$119.1	$123.8	$120.8	$123.8	$108.8	$129.8	$120.8	$123.8	$117.0	$130.4	$120.1	$123.8
24	$132.7	$135.1	$119.5	$124.2	$121.2	$124.2	$109.2	$130.1	$121.2	$124.2	$117.3	$130.8	$120.5	$124.2
25	$133.1	$135.5	$119.9	$124.5	$121.6	$124.5	$109.5	$130.4	$121.6	$124.5	$117.7	$131.1	$120.9	$124.5
26	$133.6	$135.8	$120.2	$124.8	$122.0	$124.8	$109.9	$130.8	$122.0	$124.8	$118.1	$131.5	$121.3	$124.8
27	$134.0	$136.2	$120.6	$125.2	$122.4	$125.2	$110.2	$131.1	$122.4	$125.2	$118.5	$131.8	$121.7	$125.2

Table C.9—continued

YOPE	Authorizing Official (611)		Cyber Defense Analyst (511)		Cyber Defense Incident Responder (531)		Cyber Operator (321)		Security Control Assessor (612)		Software Developer (621)		Systems Security Analyst (461)	
	DoD	Private	DoD	Private	DoD	Private	DoD	Private	DoD	Private	DoD	Private	DoD	Private
28	$134.4	$136.6	$121.0	$125.5	$122.8	$125.5	$110.6	$131.5	$122.8	$125.5	$118.9	$132.2	$122.1	$125.5
29	$134.9	$136.9	$121.4	$125.8	$123.2	$125.8	$110.9	$131.8	$123.2	$125.8	$119.3	$132.5	$122.5	$125.8
30	$135.3	$137.3	$121.8	$126.2	$123.6	$126.2	$111.3	$132.2	$123.6	$126.2	$119.6	$132.9	$122.9	$126.2

NOTES: Values are predictions from a regression model that accounts for years of potential experience, education, gender, and work role using data from the 2012–2018 ACS and the September 2018 DMDC CMF. Predictions reflect 2018 male workers with a bachelor's degree. YOPE = years of potential experience.

Table C.10
Predicted Average Annual Pay for DoD Civilian and Private-Sector Cyber Work Roles in San Francisco LPA

	Authorizing Official (611)		Cyber Defense Analyst (511)		Cyber Defense Incident Responder (531)		Cyber Operator (321)		Security Control Assessor (612)		Software Developer (621)		Systems Security Analyst (461)	
YOPE	DoD	Private	DoD	Private	DoD	Private	DoD	Private	DoD	Private	DoD	Private	DoD	Private
0	$73.3	$73.7	$66.0	$67.7	$67.0	$67.7	$60.3	$71.0	$67.0	$67.7	$64.9	$71.4	$66.6	$67.7
1	$77.3	$79.2	$69.6	$72.8	$70.6	$72.8	$63.6	$76.3	$70.6	$72.8	$68.4	$76.7	$70.2	$72.8
2	$81.5	$85.2	$73.3	$78.3	$74.4	$78.3	$67.0	$82.0	$74.4	$78.3	$72.0	$82.4	$74.0	$78.3
3	$85.9	$91.5	$77.3	$84.1	$78.4	$84.1	$70.6	$88.1	$78.4	$84.1	$75.9	$88.6	$78.0	$84.1
4	$90.5	$98.4	$81.5	$90.4	$82.7	$90.4	$74.4	$94.7	$82.6	$90.4	$80.0	$95.2	$82.2	$90.4
5	$95.4	$105.8	$85.9	$97.2	$87.1	$97.2	$78.5	$101.8	$87.1	$97.2	$84.3	$102.3	$86.6	$97.2
6	$98.7	$109.7	$88.9	$100.8	$90.2	$100.8	$81.2	$105.6	$90.1	$100.8	$87.3	$106.2	$89.6	$100.8
7	$102.2	$113.8	$92.0	$104.6	$93.4	$104.6	$84.1	$109.6	$93.3	$104.6	$90.4	$110.1	$92.8	$104.6
8	$105.8	$118.0	$95.2	$108.5	$96.6	$108.5	$87.0	$113.7	$96.6	$108.5	$93.5	$114.2	$96.0	$108.5
9	$109.5	$122.5	$98.6	$112.5	$100.0	$112.5	$90.1	$117.9	$100.0	$112.5	$96.8	$118.5	$99.4	$112.5
10	$113.3	$127.0	$102.0	$116.7	$103.5	$116.7	$93.2	$122.3	$103.5	$116.7	$100.2	$122.9	$102.9	$116.7
11	$115.3	$131.2	$103.8	$120.5	$105.4	$120.5	$94.9	$126.3	$105.3	$120.5	$102.0	$127.0	$104.7	$120.5
12	$117.4	$135.5	$105.7	$124.5	$107.2	$124.5	$96.6	$130.4	$107.2	$124.5	$103.8	$131.1	$106.6	$124.5
13	$119.5	$139.9	$107.6	$128.6	$109.2	$128.6	$98.3	$134.7	$109.1	$128.6	$105.6	$135.4	$108.5	$128.6

Table C.10—continued

	Authorizing Official (611)		Cyber Defense Analyst (511)		Cyber Defense Incident Responder (531)		Cyber Operator (321)		Security Control Assessor (612)		Software Developer (621)		Systems Security Analyst (461)	
YOPE	DoD	Private	DoD	Private	DoD	Private	DoD	Private	DoD	Private	DoD	Private	DoD	Private
14	$121.6	$144.5	$109.5	$132.8	$111.1	$132.8	$100.0	$139.1	$111.0	$132.8	$107.5	$139.8	$110.4	$132.8
15	$123.8	$149.2	$111.4	$137.1	$113.1	$137.1	$101.8	$143.7	$113.0	$137.1	$109.4	$144.4	$112.4	$137.1
16	$125.2	$152.9	$112.7	$140.5	$114.4	$140.5	$103.0	$147.2	$114.3	$140.5	$110.7	$148.0	$113.7	$140.5
17	$126.6	$156.7	$114.0	$144.0	$115.7	$144.0	$104.2	$150.9	$115.7	$144.0	$112.0	$151.7	$115.0	$144.0
18	$128.1	$160.6	$115.3	$147.6	$117.0	$147.6	$105.4	$154.6	$117.0	$147.6	$113.3	$155.4	$116.3	$147.6
19	$129.6	$164.6	$116.7	$151.2	$118.4	$151.2	$106.6	$158.5	$118.3	$151.2	$114.6	$159.3	$117.7	$151.2
20	$131.1	$168.7	$118.0	$155.0	$119.8	$155.0	$107.8	$162.4	$119.7	$155.0	$115.9	$163.2	$119.0	$155.0
21	$131.5	$169.1	$118.4	$155.4	$120.2	$155.4	$108.2	$162.9	$120.1	$155.4	$116.3	$163.7	$119.4	$155.4
22	$131.9	$169.6	$118.8	$155.8	$120.5	$155.8	$108.5	$163.3	$120.5	$155.8	$116.7	$164.1	$119.8	$155.8
23	$132.4	$170.0	$119.2	$156.2	$120.9	$156.2	$108.9	$163.7	$120.9	$156.2	$117.0	$164.6	$120.2	$156.2
24	$132.8	$170.5	$119.6	$156.7	$121.3	$156.7	$109.2	$164.2	$121.3	$156.7	$117.4	$165.0	$120.6	$156.7
25	$133.2	$170.9	$119.9	$157.1	$121.7	$157.1	$109.6	$164.6	$121.7	$157.1	$117.8	$165.4	$121.0	$157.1
26	$133.7	$171.4	$120.3	$157.5	$122.1	$157.5	$110.0	$165.0	$122.1	$157.5	$118.2	$165.9	$121.4	$157.5
27	$134.1	$171.9	$120.7	$157.9	$122.5	$157.9	$110.3	$165.5	$122.5	$157.9	$118.6	$166.3	$121.8	$157.9

Table C.10—continued

	Authorizing Official (611)		Cyber Defense Analyst (511)		Cyber Defense Incident Responder (531)		Cyber Operator (321)		Security Control Assessor (612)		Software Developer (621)		Systems Security Analyst (461)	
YOPE	DoD	Private	DoD	Private	DoD	Private	DoD	Private	DoD	Private	DoD	Private	DoD	Private
28	$134.5	$172.3	$121.1	$158.3	$122.9	$158.3	$110.7	$165.9	$122.9	$158.3	$119.0	$166.8	$122.2	$158.3
29	$135.0	$172.8	$121.5	$158.8	$123.3	$158.8	$111.0	$166.4	$123.2	$158.8	$119.3	$167.2	$122.6	$158.8
30	$135.4	$173.2	$121.9	$159.2	$123.7	$159.2	$111.4	$166.8	$123.6	$159.2	$119.7	$167.7	$123.0	$159.2

NOTES: Values are predictions from a regression model that accounts for years of potential experience, education, gender, and work role using data from the 2012–2018 ACS and the September 2018 DMDC CMF. Predictions reflect 2018 male workers with a bachelor's degree.
YOPE = years of potential experience.

Table C.11
Predicted Average Annual Pay for DoD Civilian and Private-Sector Cyber Work Roles in Seattle LPA

	Authorizing Official (611)		Cyber Defense Analyst (511)		Cyber Defense Incident Responder (531)		Cyber Operator (321)		Security Control Assessor (612)		Software Developer (621)		Systems Security Analyst (461)	
YOPE	DoD	Private	DoD	Private	DoD	Private	DoD	Private	DoD	Private	DoD	Private	DoD	Private
0	$51.6	$68.3	$46.4	$62.8	$47.1	$62.8	$42.4	$65.8	$47.1	$62.8	$45.6	$66.1	$46.8	$62.8
1	$54.3	$73.4	$48.9	$67.5	$49.7	$67.5	$44.7	$70.7	$49.6	$67.5	$48.1	$71.1	$49.4	$67.5
2	$57.3	$78.9	$51.6	$72.5	$52.3	$72.5	$47.1	$76.0	$52.3	$72.5	$50.6	$76.4	$52.0	$72.5
3	$60.4	$84.8	$54.3	$78.0	$55.2	$78.0	$49.7	$81.7	$55.1	$78.0	$53.4	$82.1	$54.8	$78.0
4	$63.6	$91.2	$57.3	$83.8	$58.1	$83.8	$52.3	$87.8	$58.1	$83.8	$56.3	$88.3	$57.8	$83.8
5	$67.1	$98.0	$60.4	$90.1	$61.3	$90.1	$55.2	$94.4	$61.2	$90.1	$59.3	$94.9	$60.9	$90.1
6	$69.4	$101.7	$62.5	$93.4	$63.4	$93.4	$57.1	$97.9	$63.4	$93.4	$61.4	$98.4	$63.0	$93.4
7	$71.8	$105.5	$64.7	$96.9	$65.6	$96.9	$59.1	$101.6	$65.6	$96.9	$63.5	$102.1	$65.2	$96.9
8	$74.4	$109.4	$66.9	$100.5	$67.9	$100.5	$61.2	$105.4	$67.9	$100.5	$65.8	$105.9	$67.5	$100.5
9	$77.0	$113.5	$69.3	$104.3	$70.3	$104.3	$63.3	$109.3	$70.3	$104.3	$68.1	$109.8	$69.9	$104.3
10	$79.7	$117.7	$71.7	$108.2	$72.8	$108.2	$65.5	$113.4	$72.8	$108.2	$70.4	$113.9	$72.3	$108.2
11	$81.1	$121.6	$73.0	$111.7	$74.1	$111.7	$66.7	$117.1	$74.0	$111.7	$71.7	$117.7	$73.6	$111.7
12	$82.5	$125.6	$74.3	$115.4	$75.4	$115.4	$67.9	$120.9	$75.4	$115.4	$73.0	$121.5	$74.9	$115.4
13	$84.0	$129.7	$75.6	$119.2	$76.7	$119.2	$69.1	$124.9	$76.7	$119.2	$74.3	$125.5	$76.3	$119.2

Table C.11—continued

	Authorizing Official (611)		Cyber Defense Analyst (511)		Cyber Defense Incident Responder (531)		Cyber Operator (321)		Security Control Assessor (612)		Software Developer (621)		Systems Security Analyst (461)	
YOPE	DoD	Private	DoD	Private	DoD	Private	DoD	Private	DoD	Private	DoD	Private	DoD	Private
14	$85.5	$133.9	$77.0	$123.0	$78.1	$123.0	$70.3	$128.9	$78.1	$123.0	$75.6	$129.6	$77.6	$123.0
15	$87.0	$138.3	$78.3	$127.1	$79.5	$127.1	$71.6	$133.2	$79.5	$127.1	$76.9	$133.8	$79.0	$127.1
16	$88.0	$141.7	$79.2	$130.2	$80.4	$130.2	$72.4	$136.5	$80.4	$130.2	$77.8	$137.2	$79.9	$130.2
17	$89.0	$145.2	$80.2	$133.5	$81.4	$133.5	$73.2	$139.9	$81.3	$133.5	$78.7	$140.6	$80.9	$133.5
18	$90.1	$148.9	$81.1	$136.8	$82.3	$136.8	$74.1	$143.3	$82.3	$136.8	$79.6	$144.1	$81.8	$136.8
19	$91.1	$152.5	$82.0	$140.2	$83.2	$140.2	$74.9	$146.9	$83.2	$140.2	$80.6	$147.6	$82.7	$140.2
20	$92.2	$156.3	$83.0	$143.7	$84.2	$143.7	$75.8	$150.5	$84.2	$143.7	$81.5	$151.3	$83.7	$143.7
21	$92.5	$156.8	$83.2	$144.0	$84.5	$144.0	$76.1	$150.9	$84.4	$144.0	$81.8	$151.7	$84.0	$144.0
22	$92.8	$157.2	$83.5	$144.4	$84.8	$144.4	$76.3	$151.3	$84.7	$144.4	$82.0	$152.1	$84.2	$144.4
23	$93.1	$157.6	$83.8	$144.8	$85.0	$144.8	$76.6	$151.7	$85.0	$144.8	$82.3	$152.5	$84.5	$144.8
24	$93.4	$158.0	$84.1	$145.2	$85.3	$145.2	$76.8	$152.2	$85.3	$145.2	$82.6	$152.9	$84.8	$145.2
25	$93.7	$158.4	$84.3	$145.6	$85.6	$145.6	$77.1	$152.6	$85.5	$145.6	$82.8	$153.3	$85.1	$145.6
26	$94.0	$158.9	$84.6	$146.0	$85.9	$146.0	$77.3	$153.0	$85.8	$146.0	$83.1	$153.7	$85.3	$146.0
27	$94.3	$159.3	$84.9	$146.4	$86.1	$146.4	$77.6	$153.4	$86.1	$146.4	$83.4	$154.2	$85.6	$146.4

Table C.11—continued

	Authorizing Official (611)		Cyber Defense Analyst (511)		Cyber Defense Incident Responder (531)		Cyber Operator (321)		Security Control Assessor (612)		Software Developer (621)		Systems Security Analyst (461)	
YOPE	DoD	Private	DoD	Private	DoD	Private	DoD	Private	DoD	Private	DoD	Private	DoD	Private
28	$94.6	$159.7	$85.2	$146.8	$86.4	$146.8	$77.8	$153.8	$86.4	$146.8	$83.6	$154.6	$85.9	$146.8
29	$94.9	$160.1	$85.4	$147.2	$86.7	$147.2	$78.1	$154.2	$86.7	$147.2	$83.9	$155.0	$86.2	$147.2
30	$95.2	$160.6	$85.7	$147.5	$87.0	$147.5	$78.3	$154.6	$86.9	$147.5	$84.2	$155.4	$86.5	$147.5

NOTES: Values are predictions from a regression model that accounts for years of potential experience, education, gender, and work role using data from the 2012–2018 ACS and the September 2018 DMDC CMF. Predictions reflect 2018 male workers with a bachelor's degree. YOPE = years of potential experience.

Table C.12
Predicted Average Annual Pay for DoD Civilian and Private-Sector Cyber Work Roles in St. Louis LPA

	Authorizing Official (611)		Cyber Defense Analyst (511)		Cyber Defense Incident Responder (531)		Cyber Operator (321)		Security Control Assessor (612)		Software Developer (621)		Systems Security Analyst (461)	
YOPE	DoD	Private	DoD	Private	DoD	Private	DoD	Private	DoD	Private	DoD	Private	DoD	Private
0	$63.8	$52.1	$57.4	$47.9	$58.3	$47.9	$52.5	$50.2	$58.3	$47.9	$56.4	$50.4	$57.9	$47.9
1	$67.3	$56.0	$60.5	$51.5	$61.4	$51.5	$55.3	$53.9	$61.4	$51.5	$59.5	$54.2	$61.1	$51.5
2	$70.9	$60.2	$63.8	$55.3	$64.8	$55.3	$58.3	$58.0	$64.7	$55.3	$62.7	$58.3	$64.4	$55.3
3	$74.7	$64.7	$67.2	$59.5	$68.2	$59.5	$61.4	$62.3	$68.2	$59.5	$66.1	$62.6	$67.8	$59.5
4	$78.7	$69.6	$70.9	$63.9	$71.9	$63.9	$64.8	$67.0	$71.9	$63.9	$69.6	$67.3	$71.5	$63.9
5	$83.0	$74.8	$74.7	$68.7	$75.8	$68.7	$68.3	$72.0	$75.8	$68.7	$73.4	$72.3	$75.3	$68.7
6	$85.9	$77.5	$77.3	$71.3	$78.5	$71.3	$70.6	$74.7	$78.4	$71.3	$75.9	$75.0	$78.0	$71.3
7	$88.9	$80.4	$80.0	$73.9	$81.2	$73.9	$73.1	$77.5	$81.2	$73.9	$78.6	$77.8	$80.7	$73.9
8	$92.0	$83.4	$82.8	$76.7	$84.1	$76.7	$75.7	$80.3	$84.0	$76.7	$81.4	$80.8	$83.6	$76.7
9	$95.2	$86.6	$85.7	$79.5	$87.0	$79.5	$78.3	$83.3	$87.0	$79.5	$84.2	$83.8	$86.5	$79.5
10	$98.6	$89.8	$88.7	$82.5	$90.1	$82.5	$81.1	$86.5	$90.0	$82.5	$87.2	$86.9	$89.5	$82.5
11	$100.3	$92.7	$90.3	$85.2	$91.7	$85.2	$82.5	$89.3	$91.6	$85.2	$88.7	$89.7	$91.1	$85.2
12	$102.1	$95.8	$91.9	$88.0	$93.3	$88.0	$84.0	$92.2	$93.3	$88.0	$90.3	$92.7	$92.7	$88.0
13	$103.9	$98.9	$93.6	$90.9	$95.0	$90.9	$85.5	$95.2	$94.9	$90.9	$91.9	$95.7	$94.4	$90.9

Table C.12—continued

	Authorizing Official (611)		Cyber Defense Analyst (511)		Cyber Defense Incident Responder (531)		Cyber Operator (321)		Security Control Assessor (612)		Software Developer (621)		Systems Security Analyst (461)	
YOPE	DoD	Private	DoD	Private	DoD	Private	DoD	Private	DoD	Private	DoD	Private	DoD	Private
14	$105.8	$102.1	$95.2	$93.8	$96.7	$93.8	$87.0	$98.3	$96.6	$93.8	$93.5	$98.8	$96.1	$93.8
15	$107.7	$105.5	$96.9	$96.9	$98.4	$96.9	$88.6	$101.6	$98.3	$96.9	$95.2	$102.1	$97.8	$96.9
16	$108.9	$108.1	$98.1	$99.3	$99.5	$99.3	$89.6	$104.1	$99.5	$99.3	$96.3	$104.6	$98.9	$99.3
17	$110.2	$110.8	$99.2	$101.8	$100.7	$101.8	$90.6	$106.7	$100.6	$101.8	$97.4	$107.2	$100.1	$101.8
18	$111.5	$113.5	$100.3	$104.3	$101.8	$104.3	$91.7	$109.3	$101.8	$104.3	$98.6	$109.9	$101.2	$104.3
19	$112.7	$116.3	$101.5	$106.9	$103.0	$106.9	$92.7	$112.0	$103.0	$106.9	$99.7	$112.6	$102.4	$106.9
20	$114.0	$119.2	$102.7	$109.6	$104.2	$109.6	$93.8	$114.8	$104.1	$109.6	$100.8	$115.4	$103.6	$109.6
21	$114.4	$119.5	$103.0	$109.9	$104.5	$109.9	$94.1	$115.1	$104.5	$109.9	$101.2	$115.7	$103.9	$109.9
22	$114.8	$119.9	$103.3	$110.1	$104.9	$110.1	$94.4	$115.4	$104.8	$110.1	$101.5	$116.0	$104.2	$110.1
23	$115.2	$120.2	$103.7	$110.4	$105.2	$110.4	$94.7	$115.7	$105.2	$110.4	$101.8	$116.3	$104.6	$110.4
24	$115.5	$120.5	$104.0	$110.7	$105.6	$110.7	$95.0	$116.0	$105.5	$110.7	$102.2	$116.6	$104.9	$110.7
25	$115.9	$120.8	$104.3	$111.0	$105.9	$111.0	$95.3	$116.4	$105.8	$111.0	$102.5	$116.9	$105.3	$111.0
26	$116.3	$121.2	$104.7	$111.3	$106.2	$111.3	$95.7	$116.7	$106.2	$111.3	$102.8	$117.3	$105.6	$111.3
27	$116.7	$121.5	$105.0	$111.6	$106.6	$111.6	$96.0	$117.0	$106.5	$111.6	$103.2	$117.6	$105.9	$111.6

Table C.12—continued

	Authorizing Official (611)		Cyber Defense Analyst (511)		Cyber Defense Incident Responder (531)		Cyber Operator (321)		Security Control Assessor (612)		Software Developer (621)		Systems Security Analyst (461)	
YOPE	DoD	Private	DoD	Private	DoD	Private	DoD	Private	DoD	Private	DoD	Private	DoD	Private
28	$117.0	$121.8	$105.4	$111.9	$106.9	$111.9	$96.3	$117.3	$106.9	$111.9	$103.5	$117.9	$106.3	$111.9
29	$117.4	$122.1	$105.7	$112.2	$107.3	$112.2	$96.6	$117.6	$107.2	$112.2	$103.8	$118.2	$106.6	$112.2
30	$117.8	$122.5	$106.1	$112.5	$107.6	$112.5	$96.9	$117.9	$107.6	$112.5	$104.2	$118.5	$107.0	$112.5

NOTES: Values are predictions from a regression model that accounts for years of potential experience, education, gender, and work role using data from the 2012–2018 ACS and the September 2018 DMDC CMF. Predictions reflect 2018 male workers with a bachelor's degree.
YOPE = years of potential experience.

Table C.13
Predicted Average Annual Pay for DoD Civilian and Private-Sector Cyber Work Roles in Tucson LPA

	Authorizing Official (611)		Cyber Defense Analyst (511)		Cyber Defense Incident Responder (531)		Cyber Operator (321)		Security Control Assessor (612)		Software Developer (621)		Systems Security Analyst (461)	
YOPE	DoD	Private	DoD	Private	DoD	Private	DoD	Private	DoD	Private	DoD	Private	DoD	Private
0	$63.4	$45.9	$57.1	$42.2	$57.9	$42.2	$52.2	$44.2	$57.9	$42.2	$56.1	$44.4	$57.6	$42.2
1	$66.8	$49.3	$60.2	$45.3	$61.0	$45.3	$55.0	$47.5	$61.0	$45.3	$59.1	$47.7	$60.7	$45.3
2	$70.4	$53.0	$63.4	$48.7	$64.3	$48.7	$57.9	$51.1	$64.3	$48.7	$62.3	$51.3	$64.0	$48.7
3	$74.2	$57.0	$66.8	$52.4	$67.8	$52.4	$61.1	$54.9	$67.8	$52.4	$65.6	$55.2	$67.4	$52.4
4	$78.2	$61.3	$70.4	$56.3	$71.5	$56.3	$64.3	$59.0	$71.4	$56.3	$69.2	$59.3	$71.0	$56.3
5	$82.4	$65.8	$74.2	$60.5	$75.3	$60.5	$67.8	$63.4	$75.3	$60.5	$72.9	$63.7	$74.9	$60.5
6	$85.3	$68.3	$76.8	$62.8	$78.0	$62.8	$70.2	$65.8	$77.9	$62.8	$75.5	$66.1	$77.5	$62.8
7	$88.3	$70.8	$79.5	$65.1	$80.7	$65.1	$72.7	$68.2	$80.7	$65.1	$78.1	$68.6	$80.2	$65.1
8	$91.4	$73.5	$82.3	$67.5	$83.5	$67.5	$75.2	$70.8	$83.5	$67.5	$80.8	$71.1	$83.0	$67.5
9	$94.6	$76.2	$85.2	$70.1	$86.5	$70.1	$77.8	$73.4	$86.4	$70.1	$83.7	$73.8	$85.9	$70.1
10	$98.0	$79.1	$88.2	$72.7	$89.5	$72.7	$80.6	$76.2	$89.4	$72.7	$86.6	$76.5	$89.0	$72.7
11	$99.7	$81.7	$89.8	$75.0	$91.1	$75.0	$82.0	$78.6	$91.0	$75.0	$88.2	$79.0	$90.5	$75.0
12	$101.5	$84.3	$91.4	$77.5	$92.7	$77.5	$83.5	$81.2	$92.7	$77.5	$89.7	$81.6	$92.1	$77.5

Table C.13—continued

	Authorizing Official (611)		Cyber Defense Analyst (511)		Cyber Defense Incident Responder (531)		Cyber Operator (321)		Security Control Assessor (612)		Software Developer (621)		Systems Security Analyst (461)	
YOPE	DoD	Private	DoD	Private	DoD	Private	DoD	Private	DoD	Private	DoD	Private	DoD	Private
13	$103.3	$87.1	$93.0	$80.0	$94.4	$80.0	$85.0	$83.9	$94.3	$80.0	$91.3	$84.3	$93.8	$80.0
14	$105.1	$90.0	$94.6	$82.7	$96.0	$82.7	$86.5	$86.6	$96.0	$82.7	$93.0	$87.1	$95.5	$82.7
15	$107.0	$92.9	$96.3	$85.4	$97.8	$85.4	$88.0	$89.4	$97.7	$85.4	$94.6	$89.9	$97.2	$85.4
16	$108.2	$95.2	$97.4	$87.5	$98.9	$87.5	$89.0	$91.7	$98.8	$87.5	$95.7	$92.1	$98.3	$87.5
17	$109.5	$97.6	$98.6	$89.7	$100.0	$89.7	$90.1	$93.9	$100.0	$89.7	$96.8	$94.4	$99.4	$89.7
18	$110.7	$100.0	$99.7	$91.9	$101.2	$91.9	$91.1	$96.3	$101.1	$91.9	$97.9	$96.8	$100.6	$91.9
19	$112.0	$102.5	$100.9	$94.2	$102.3	$94.2	$92.1	$98.7	$102.3	$94.2	$99.1	$99.2	$101.7	$94.2
20	$113.3	$105.0	$102.0	$96.5	$103.5	$96.5	$93.2	$101.1	$103.5	$96.5	$100.2	$101.6	$102.9	$96.5
21	$113.7	$105.3	$102.3	$96.8	$103.9	$96.8	$93.5	$101.4	$103.8	$96.8	$100.5	$101.9	$103.2	$96.8
22	$114.1	$105.6	$102.7	$97.0	$104.2	$97.0	$93.8	$101.7	$104.1	$97.0	$100.9	$102.2	$103.6	$97.0
23	$114.4	$105.9	$103.0	$97.3	$104.5	$97.3	$94.1	$101.9	$104.5	$97.3	$101.2	$102.5	$103.9	$97.3
24	$114.8	$106.1	$103.3	$97.5	$104.9	$97.5	$94.4	$102.2	$104.8	$97.5	$101.5	$102.7	$104.2	$97.5
25	$115.2	$106.4	$103.7	$97.8	$105.2	$97.8	$94.7	$102.5	$105.2	$97.8	$101.8	$103.0	$104.6	$97.8
26	$115.5	$106.7	$104.0	$98.1	$105.6	$98.1	$95.0	$102.8	$105.5	$98.1	$102.2	$103.3	$104.9	$98.1

Table C.13—continued

YOPE	Authorizing Official (611)		Cyber Defense Analyst (511)		Cyber Defense Incident Responder (531)		Cyber Operator (321)		Security Control Assessor (612)		Software Developer (621)		Systems Security Analyst (461)	
	DoD	Private	DoD	Private	DoD	Private	DoD	Private	DoD	Private	DoD	Private	DoD	Private
27	$115.9	$107.0	$104.4	$98.3	$105.9	$98.3	$95.4	$103.0	$105.9	$98.3	$102.5	$103.6	$105.3	$98.3
28	$116.3	$107.3	$104.7	$98.6	$106.2	$98.6	$95.7	$103.3	$106.2	$98.6	$102.8	$103.8	$105.6	$98.6
29	$116.7	$107.6	$105.0	$98.8	$106.6	$98.8	$96.0	$103.6	$106.5	$98.8	$103.2	$104.1	$105.9	$98.8
30	$117.0	$107.9	$105.4	$99.1	$106.9	$99.1	$96.3	$103.9	$106.9	$99.1	$103.5	$104.4	$106.3	$99.1

NOTES: Values are predictions from a regression model that accounts for years of potential experience, education, gender, and work role using data from the 2012–2018 ACS and the September 2018 DMDC CMF. Predictions reflect 2018 male workers with a bachelor's degree.